CABI Regional Tourism Series

by

Konstantinos Andriotis Middlesex University, London
and
Dimitrios Stylidis Middlesex University, London

Tourism literature includes a plethora of books and conference and journal papers that address various issues of tourism in different regions and sub-regions of the world. However, up to now there has been a lack of a collection of contribution-based books with a regional and sub-regional focus. In order to complement the existing tourism literature with new or emerging topics of particular interest, it is the aim of the **CABI Regional Tourism Series** to offer a collection of books with comparative and individual case studies from at least eight regions and sub-regions of the world. These books will come under the following titles, each of which focuses on some of the countries or country clusters: Tourism in Western Europe; Tourism in Eastern Europe; Tourism in East and Southeast Asia; Tourism in Middle East; Tourism in Sub-Saharan Africa; Tourism in the Caribbean; Tourism in Latin America; Tourism in Pacific Oceania.

Titles Available:

Tourism Planning and Development in Latin America Carlos Monterrubio, Konstantinos Andriotis and Dimitrios Stylidis
Tourism Planning and Development in South Asia Dimitrios Stylidis and Boopen Seetanah
Tourism Marketing in Western Europe Nikolaos Boukas and Dimitrios Stylidis

Titles Forthcoming:

Tourism Planning and Development in the Middle East Stella Kladou, Anna Farmaki, Konstantinos Andriotis and Dimitrios Stylidis
Tourism Development and Planning in Western Europe Konstantinos Andriotis, Carla Pinto Cardoso and Dimitrios Stylidis

Suggested Topics:

- Tourism Development and Planning
- Tourism Economics and Finance
- Tourism Marketing, Management and Information Technology

To discuss a proposal for a book in this series, please contact Konstantinos Andriotis at k.andriotis@mdx.ac.uk or Dimitrios Stylidis at d.stylidis@mdx.ac.uk.

The Series Editors

Konstantinos Andriotis is Professor in Tourism at Middlesex University London and in the past he was Associate Professor and Head of the Hotel and Tourism Management Department at the Cyprus University of Technology. He holds an MSc in International Hospitality Management (University of Strathclyde, 1995), a PhD in Tourism Development and Planning (Bournemouth University, 2000) and a Postdoctorate in Tourism Marketing (Hellenic Open University, 2006). He edits the *International Journal of Tourism Policy* and the *Journal of Qualitative Research in Tourism*. He is a member of the Editorial Board of more than 20 journals and Associate Editor of the leading tourism journal *Annals of Tourism Research*. He is author of seven books and over 45 peer-reviewed articles and book chapters. His work has appeared in leading academic journals, such as *Annals of Tourism Research, Journal of Travel Research, Tourism Management* and *Journal of Sustainable Tourism*. He has an h-index of 29 and more than 3700 citations. He has acted as Chairman and member of the organizing and scientific committees of various international conferences and has given keynote speeches at several conferences. His research interests are in tourism development and planning, alternative forms of tourism, tourism experience and degrowth.

Dimitrios Stylidis is Senior Lecturer in Tourism at Middlesex University London. Prior to joining Middlesex University, he worked for two years as a Postdoctoral Researcher at Ben-Gurion University of the Negev, Israel. He holds a PhD and an MSc in Tourism Management from the University of Surrey, UK. He is the research notes editor of the *International Journal of Tourism Policy*. His work has appeared in leading academic journals including the *Annals of Tourism Research, Journal of Travel Research* and *Tourism Management*. His current research interests include place image, tourism impacts and residents' attitudes toward tourism development.

Tourism Marketing in Western Europe

Edited by

Nikolaos Boukas
European University Cyprus, Cyprus

and

Dimitrios Stylidis
Middlesex University London, UK

CABI

CABI is a trading name of CAB International

CABI
Nosworthy Way
Wallingford
Oxfordshire OX10 8DE
UK

Tel: +44 (0)1491 832111
E-mail: info@cabi.org
Website: www.cabi.org

CABI
WeWork
One Lincoln St
24th Floor
Boston, MA 02111
USA

T: +1 (617)682-9015
E-mail: cabi-nao@cabi.org

A catalogue record for this book is available from the British Library, London, UK.

Library of Congress Control Number: 2021939511

References to internet websites (URLs) were accurate at the time of writing.

ISBN: 9781789248753 (hardback)
9781789248760 (ePDF)
9781789248777 (ePub)

DOI: 10.1079/9781789248753.0000

Commissioning Editor: Claire Parfitt
Editorial Assistant: Ali Thompson
Production Editor: Tim Kapp

Typeset by Exeter Premedia Services Pvt Ltd, Chennai, India
Printed and bound in the UK by Severn, Gloucester

Contents

Contributors

Jordi Arcos-Pumarola holds a PhD in Education and Society from the University of Lleida. He is director of the Research Group on Tourism, Culture and Territory at Barcelona School of Tourism, Hospitality and Gastronomy CETT (Universitat de Barcelona), where he also teaches Literary and Cultural Tourism and Tourism Research Methods. His academic research focuses on cultural and literary tourism and tourism ethics.

CETT-UB, Barcelona School of Tourism, Hospitality and Gastronomy – University of Barcelona, Avinguda Can Marcet 36–38, 08035 Barcelona, Spain. E-mail: jordi.arcos@cett.cat

Thomas Bausch is Professor and Director of the Competence Centre Tourism and Mobility at Free University of Bozen-Bolzano. His research focuses on destination development, sustainable regional development and tourism mobility, with a special emphasis on Alpine destinations.

Competence Centre Tourism and Mobility, Free University of Bozen-Bolzano, Piazzetta dell'Università 1, I - 39031 Brunico, Italy. E-mail: thomas.bausch@unibz.it

Bernardo Borges graduated in Marketing at the University of Aveiro in 2017, completing his Master's degree in Tourism Management and Planning at the same university in 2019. In 2020 he completed an intensive digital marketing course at EDIT Porto. He currently works as a consultant at IDTOUR, a tourism consultant company based in Aveiro, participating in numerous management and marketing tourism projects and devising communication strategies applied to many different scenarios.

Department of Economics, Management, Industrial Engineering and Tourism, University of Aveiro, Campus Universitário de Santiago, 3810-193 Aveiro, Portugal. E-mail: b.borges@ua.pt

Nikolaos Boukas holds a PhD in Management Studies from the University of Exeter, UK. He is currently an Associate Professor of Tourism and Hospitality Management and the Director of the Center for Sustainable Management of Tourism, Sport and Events (CESMATSE) of the European University Cyprus. His research interests are focused on tourism marketing, tourism planning and sustainable development, tourism policy, niche tourism, island tourism, cultural heritage tourism, and rural tourism.

Center for Sustainable Management of Tourism, Sport & Events (CESMATSE), European University Cyprus, 6 Diogenes Street, Engomi, PO Box 22006, 1516 Nicosia, Cyprus. E-mail: n.boukas@euc.ac.cy

Francesco Calza is the Vice Dean of the Parthenope University of Naples, where he has taught Corporate Strategy, Management and International Finance for many years. He was also Professor in International Finance at the Wirtshaft University of Vienna and taught on many MBA programmes in Italy. He has studied growth strategies of firms in international markets, logistics and energy management and has authored many studies about management and cross-cultural management, with a specific focus on the internationalization process of firms from Mediterranean countries. He is an editorial board member for various journals in management.

Department of Management and Quantitative Studies, Parthenope University of Naples, Via G. Parisi, 13 - 80132 Naples, Italy. E-mail: francesco.calza@uniparthenope.it

Jacey Choe is Assistant Professor in the Department of Integrated Resort and Tourism Management at the Faculty of Business Administration, University of Macau. She received her PhD in the School of Hotel and Tourism Management at the Hong Kong Polytechnic University. Her research interests are psychological and cultural aspects of food, food technology and culinary tourism.

Department of Integrated Resort and Tourism Management, Faculty of Business Administration, University of Macau, Avenida da Universidade, Taipa, Macau, China. E-mail: jaceychoe@um.edu.mo

Marta Conill-Tetuà holds a PhD in Didactics of Heritage, Arts and Cultural Tourism from the University of Barcelona and a Master's degree in Innovation in Tourism Management. She is currently working for the Vallès Occidental county council on the tourist management of the territory, and is particularly involved in the gastronomy tourism project "Xarxa de Productes de la Terra". Her main research interests are cultural tourism and cultural heritage mediation.

Consell Comarcal del Vallès Occidental, Terrassa, Spain. E-mail: conilltm@ccvoc.cat

Elizabeth Cooper is an Industrial PhD Fellow at Copenhagen Business School and Visit Greenland, Greenland's national DMO. Her PhD research aims to link the fields of behavioural economics and tourism, in order to encourage more sustainable behaviour among cruise tourists. Her current research interests include: experimental research in tourism, marketing of polar and peripheral destinations, scientific tourism and tourist behaviour.

Department of Management, Society & Communication, Copenhagen Business School, Dalgas Have 15, 2000 Frederiksberg, Denmark. E-mail: eco.msc@cbs.dk

Rui Augusto Costa holds a PhD in Tourism at the University of Aveiro. He is Assistant Professor and member of the Executive Board in the Department of Economics, Management, Industrial Engineering and Tourism. He is also an Integrated Member of the Research Unit on Governance, Competitiveness and Public Policy in the Research Group on Tourism and Development, Associate Editor of the *Journal of Tourism & Development* and Internship Coordinator (Bachelor and Master). His research interests are in: tourism planning, tourism marketing, networks, governance and public policy, territorial dynamics of investment and financing of small and micro-enterprises in the tourism sector.

Department of Economics, Management, Industrial Engineering and Tourism, University of Aveiro, Campus Universitário de Santiago, 3810-193 Aveiro, Portugal. E-mail: rui.costa@ua.pt

Mengyun Hu is a PhD student at the School of Management, University of Bristol, UK. She has a Master's degree in marketing. Her research projects focus on marketing and tourism technology.

School of Management, Howard House, Queens Avenue, University of Bristol, Bristol BS8 1SD, UK. E-mail: mengyun.hu@bristol.ac.uk

Elitza Iordanova holds a PhD from Brunel University, UK, and a Magister/Magistra rerum socialium economicarumque degree in International Business Administration from the Vienna University of Economics and Business Administration. She is Head of Subject for Tourism, Aviation and Events, London Geller College of Hospitality and Tourism, University of West London. Her research interests are mainly in destination marketing and consumer behaviour.

University of West London, London Geller College of Hospitality and Tourism, St Mary's Road, Ealing, London W5 5RF, UK. E-mail: elitza.Iordanova@uwl.ac.uk

Gil Jannes has a PhD in Theoretical Physics and currently works as an Assistant Professor of Statistics at the Faculty of Economics & Business of the Complutense University in Madrid, Spain. He has published more than 20 research articles about Hawking radiation and other aspects of black hole physics. His current research interests include the

application of quantitative methods to study the economic impact of tourism in Spain.

Department of Financial and Actuarial Economics & Statistics, Complutense University of Madrid, Campus de Somosaguas, s/n, 28223 Pozuelo de Alarcón, Madrid, Spain. E-mail: gjannes@ucm.es

Eleonora Pantano is Senior Lecturer (Associate Professor) of Marketing at the University of Bristol, UK. She has a PhD in Psychology of Programming and Artificial Intelligence (2008), an MEng in Management (2005) and a PGCHE in Teaching and Supporting Learning (2016). Her findings appear in several books and textbooks, and in international peer-reviewed journals (e.g. *Computers in Human Behaviour, Tourism Management, Journal of Business Research* and *Psychology and Marketing*). Her 80+ papers have consistently pushed the frontiers of marketing and retailing and have become highly cited by academics in the field (h-index 21, Scopus).

School of Management, Howard House, Queens Avenue, University of Bristol, Bristol BS8 1SD, UK. E-mail: e.pantano@bristol.ac.uk

Jesús Barreal Pernas has a PhD in Economics and is Assistant Professor of Statistics at the Faculty of Economics & Business in the Complutense University, Madrid. His current research focuses on different economic areas, including the agricultural marketplace, natural resources and tourism. All his studies employ a sizeable quantitative methodology such as time series, cluster analysis or panel data to analyse the socioeconomic relations between producers, costumers, citizens and policymakers.

Department of Financial and Actuarial Economics & Statistics, Complutense University of Madrid, Campus de Somosaguas, s/n, 28223 Pozuelo de Alarcón, Madrid, Spain. E-mail: jbarreal@ucm.es

Carina Ren is Associate Professor, Center for Innovation and Research in Culture and Living in the Arctic at Aalborg University, Denmark, and Head of AAU Arctic. She researches relations between tourism and other fields through the lens of cultural innovation. Current interests include culturally sensitive tourism development in Greenland and the Arctic as well as digital capacity building and cruise tourism. Carina has published in leading tourism and Arctic journals and is co-editor of several special issues and books, such as *Collaborative Research Methods in Arctic Research: Experiences from Greenland*, co-edited with Anne Merrild Hansen (Routledge, 2020).

Associate Professor, Department of Culture and Learning, Aalborg University, A.C. Meyers Vænge 15, 2450 Copenhagen SV, Denmark. E-mail: ren@hum.aau.dk

Michele Simoni is Professor of Marketing and Innovation and Business Design at the Department of Management and Quantitative Studies,

Parthenope University of Naples. He has been a visiting scholar at the China Europe International Business School (CEIBS), Shanghai, and at the University of South Wales (UNSW), Sydney. He co-ordinates the Observatory on Local Innovation Systems (SLIOB), created in 2018 by the Department of Management and Quantitative Studies as part of the Department of Excellence project funded by MIUR. He is also the director of the PhD programme in Entrepreneurship and Innovation run jointly by Parthenope University and Vanvitelli University. He is the co-ordinator of the thematic division of the Italian Society of Marketing (SIM) on Marketing of Technology and Innovation.

Department of Management and Quantitative Studies, Parthenope University of Naples, Via G. Parisi, 13 - 80132 Naples, Italy. E-mail: michele.simoni@uniparthenope.it

Annarita Sorrentino is Assistant Professor of Marketing at the Department of Management and Quantitative Studies, University of Naples Parthenope, where she teaches Customer Experience Management. She has been a visiting scholar at Middlesex University and Leeds Beckett University, UK, where she has conducted research on consumer behaviour in tourism and hospitality. Her main research interests are in consumer behaviour, tourism marketing and customer experience management.

Department of Management and Quantitative Studies, Parthenope University of Naples, Via G. Parisi, 13 - 80132 Naples, Italy. E-mail: annarita.sorrentino@uniparthenope.it

Dimitrios Stylidis is a Senior Lecturer in Tourism at Middlesex University, London. Prior to this, he was a post-doctoral researcher and visiting lecturer at Ben-Gurion University of the Negev, Israel. He holds an MSc and PhD in Tourism from the University of Surrey, UK. His current research interests include place image and marketing, tourist destination image and behaviour, tourism impacts and residents' attitudes towards tourism development. His work has been published in several journals including *Tourism Management*, *Annals of Tourism Research*, *Journal of Travel Research*, *Journal of Sustainable Tourism*, and *Journal of Travel and Tourism Marketing*.

Department of Marketing, Branding and Tourism, The Business School, Middlesex University London, The Burroughs, London, NW4 4BT, UK. E-mail: d.stylidis@mdx.ac.uk

Nikolaos Stylos is Postgraduate Research Director, and Innovation & Digitalisation Research Group Lead at the School of Management, University of Bristol, UK. He has also been appointed Honorary Professor of Hotel Management at Tainan University of Technology, Taiwan. His research projects focus on decision-making in tourism marketing, consumer behaviour, digital marketing, and environmental management. He has published in various journals including *Tourism Management, Journal of Travel Research, Psychology & Marketing, Computers in Human Behaviour, Technological Forecasting and Social Change,* and *Journal of Cleaner Production.* He has served as a professional management and technical consultant for over ten years.

School of Management, Howard House, Queens Avenue, University of Bristol, Bristol BS8 1SD, UK. E-mail: n.stylos@bristol.ac.uk

Mariapina Trunfio is Associate Professor of Economics and Business Management and Director of the Master's in Tourism & Hospitality Management at the University of Naples Parthenope, Italy. Her current research focuses on destination management and governance, local development, overtourism and sustainable development, cultural heritage and creativity, cultural diversity and entrepreneurship, tourism innovation, smart destinations, smart technologies in Covid-19, digital and social media marketing, and augmented reality, virtual reality and mixed reality in tourism and cultural heritage. She has published several papers in leading journals and monographs and has authored several book chapters and conference papers.

Department of Management and Quantitative Studies, Parthenope University of Naples, Via G. Parisi, 13 - 80132 Naples, Italy. E-mail: mariapina.trunfio@uniparthenope.it

Ilma Aulia Zaim holds a Bachelor degree in Visual Communication Design and a Master's degree in Marketing from Bandung Institute of Technology, Indonesia. She recently obtained her PhD in Tourism from Middlesex University, London. Her research interests include tourism marketing, destination image, video elicitation method and cross-culture. She has worked on some tourism-related video projects with the Government of West Java, Indonesia, as well as the United Nations International Labour Organization.

Middlesex University London, The Burroughs, London NW4 4BT, UK. E-mail: ilma.zaim@gmail.com

Introduction

Nikolaos Boukas[1]* and Dimitrios Stylidis[2]

[1]*European University Cyprus, Cyprus*
[2]*Middlesex University London, UK correspondence: n.boukas@euc.ac.cy*

Tourism is characterized by diversity, enormous growth and multi-dimensional impacts at economic, sociocultural and environmental levels. Additionally, in the current turbulent environment, tourism destinations need, on the one hand, to maintain and enhance their competitiveness in the tourism map and, on the other, to keep and protect their resources' integrity for future generations, based on sustainability premises (Ritchie and Crouch, 2003). In this respect, contemporary destinations face several challenges such as those of overtourism (Peeters *et al.*, 2021), (un)sustainable development (Kuščer and Mihalič, 2019) and service quality and perceived value for the tourist (Oriade and Schofield, 2019). Furthermore, under these environmental conditions, the current, phenomenal – and largely unknown to modern humanity – crisis of Covid-19, hit significantly the tourism sector and underlined the importance of dealing efficiently and responsibly with tourism matters (Higgins-Desbiolles, 2020). These include: paying greater attention to qualitative forms of tourism; emphasizing innovative products and processes; involving local communities in the planning process; discovering new markets for tourist experiences; and developing, distributing, pricing and promoting more effectively and efficiently destination products and services.

As such, for a destination in the post-Covid-19 era, competing at world level with other destinations with multi-faceted products and, simultaneously, being sustainable is a complex equation that needs to be solved by tourism policy makers, destination management organizations (DMOs) and tourism marketing professionals. Of course, this applies to all destinations that compete in the global tourism arena, but for traditional destinations, such as those in western Europe, which have a great history in tourism but also a past with many challenges regarding tourism planning, development, management and marketing, many of them being 'mature' in terms of tourism development, this challenge requires specialized marketing management attention. In this respect, the literature needs to shed light on how destinations in the region can conceptualize and mitigate their weaknesses as well as capitalize on their competences in order to plan, develop

*Corresponding author: n.boukas@euc.ac.cy

and manage/market tourism products that could lead them to sustainable competitiveness in the long term.

In this regard, this book aims to explore and valorize marketing paradigms utilized by various western European countries in order to manage their tourism offerings and position them in the global tourism arena. The book aims to enhance tourism literature by examining and comparing contemporary marketing tools currently used in western-European countries' case studies for understanding how tourism marketing on this side of the continent has been handled and has evolved in an era of contradictory forces such as competitiveness versus sustainability, locality versus globalization, maturity versus new product development, and traditional media versus new technologies.

The region of western Europe has gone through numerous transformations at several levels. This is why a clear categorization of the area is not so easily formulated. According to United Nations World Tourism Organization (UNWTO, 2020), there are nine countries in this region: Austria, Belgium, France, Germany, Lichtenstein, Luxembourg, Monaco, The Netherlands and Switzerland. Aside, however, from the geographic location of each country, there are also other characteristics such as economic, cultural and political that add many more countries to the region. According to the Council of Europe (cited in John, 2001, p. 36), for example, western European sub-national governmental structures are: Belgium, Denmark, Finland, France, Greece, Germany, Ireland, Italy, The Netherlands, Norway, Portugal, Spain, Sweden, Switzerland and the UK. In this respect, and for the case of this edited volume, only cases for the following countries were considered: Andorra, Austria, Belgium, Denmark, Finland, France, Germany, Greece, Iceland, Italy, Ireland, Liechtenstein, Luxembourg, Malta, Monaco, Norway, The Netherlands, Portugal, San Marino, Spain, Sweden, Switzerland and the UK.

Based on the above classifications, a certain matter can be concluded: western European tourism is quite fragmented. In this regard, there are several macro- and micro-destinations in the region, offering a plethora of natural, cultural and human resources to tourists and visitors alike. Additionally, the region encompasses some of the greatest destinations in the world in terms of volume and receipts. Characteristically, Table I.1 indicates the tourist arrivals, tourism receipts and tourism economic contribution to the European continent from the countries of western Europe in 2019.

According to Table I.1, the top nine countries in terms of arrivals (France, Spain, Italy, Germany, UK, Austria, Greece, Portugal and The Netherlands) host approximately 56% of the total tourist arrivals in Europe, contributing approximately 64% of the total tourist receipts in the continent. These figures, in combination with the contribution of each individual country's tourism to its GDP and employment, signify the important role of tourism in the region.

Nonetheless, there are two issues raised while studying these statistics: the first is that western European countries are a mosaic of landscapes, destinations, cultures, attractions and activities that attract people from all over the world. The second is that the fragmented nature of the region, with small

Table I.1 Tourist arrivals and tourism receipts. (Based on data from UNWTO, 2020 and WTTC, 2020).

Country	International tourist arrivals (000s) 2019	Arrivals – contribution to market share of Europe (%) 2019	International tourism receipts (US$ million) 2019	Receipts – contribution to market share of Europe (%) 2019	Contribution of tourism to GDP (%) 2019	Contribution to employment (000s jobs) 2019
Andorra	3042	0.4	N/A	N/A	N/A	N/A
Austria	31,884	4.3	22,942	4.0	11.8	567.7
Belgium	9343	1.3	8870	1.5	10.3	239.4
Denmark	13,285	1.8	8848	1.6	6.6	190.9
Finland	3290	0.4	3726	0.6	7.5	214.0
France	89,400*	N/A	63,801	11.1	8.5	2678.1
Germany	39,563	5.3	41,638	7.2	9.1	5668.6
Greece	31,348	4.2	20,351	3.5	20.8	846.2
Iceland	10,951	1.5	6425	1.1	22.8	44.1
Italy	64,513	8.7	49,596	8.6	13.0	3475.9
Ireland	10,951	1.5	6425	1.1	4.3	135.5
Liechtenstein	98	0.0	N/A	N/A	N/A	N/A
Malta	2753	0.4	1901	0.3	15.8	52.8
Monaco	363	0.0	N/A	N/A	N/A	N/A
Norway	5879	0.8	4968	0.9	8.0	309.6
Netherlands	20,128	2.7	18,487	3.2	5.7	791.3
Portugal	24,600	3.3	20,633	3.6	16.5	902.4
San Marino	111	0.0	N/A	N/A	N/A	N/A
Spain	83,509	11.2	79,708	13.8	14.3	2878.2
Sweden	7440	N/A	15,247	2.6	8.2	500.2

Continued

Table I.1 Continued

Country	International tourist arrivals (000s) 2019	Arrivals – contribution to market share of Europe (%) 2019	International tourism receipts (US$ million) 2019	Receipts – contribution to market share of Europe (%) 2019	Contribution of tourism to GDP (%) 2019	Contribution to employment (000s jobs) 2019
Switzerland	11,818	1.6	17,949	3.1	7.6	484.7
United Kingdom	39,418	5.3	52,721	9.1	9.0	3939.5

N/A, data not available
* data for 2018 as not available for 2019

and large countries in it, complicates the collection and analysis of proper statistics for formng a holistic view of the tourism sector in the region. Based on these characteristics, therefore, as in the whole European continent, in western Europe tourism is characterized by seasonal variances from country to country, region to region and local destination to local destination (e Silva *et al.*, 2018). e Silva *et al.* (2018) also argue that cities, islands, coastal areas and the Alps are important destinations in Europe, while strictly tourist areas such as resorts and coastal ones, due to their high dependence on tourism, tend to be more susceptible to crises and seasonality in comparison to cities. Additionally, as Costa *et al.* (2013) argue, during the last decades, many European countries have invested their resources on improving their tourism management, paying attention to new product development and facilitating the process for public–private partnerships. In this respect, the role of the European Union in leveraging tourism development, not only in specific countries but also in the whole EU region, is remarkable and acts as a catalyst for the entire region.

In this book, therefore, several cases from various countries of western Europe, in regards to their destination marketing management, are studied. This edited volume is divided into 11 chapters. These chapters mainly deal with marketing issues through practices in western European destinations, such as understanding the tourist behaviour and the travel experience, segmentation issues in tourism, stakeholder participation in tourism marketing, specialized tourism products and niche tourism, product development and innovation, DMOs and tourism channels, destination image, use of social media in promoting destinations, and digital marketing. Every unique chapter studies the above themes on specific destinations of western Europe (i.e. the Alps, Austria, Italy, France, Greece, Greenland, Portugal, Spain and United Kingdom), by scholars and professionals in tourism marketing research.

Following this introduction, the first chapter is a conceptual work by Mengyun Hu, Eleonora Pantano and Nikolaos Stylos addressing the subject of the Internet of Things (IoT) technologies and how these have impacted on the reformulation of services provided and received. In this chapter, the authors suggest that the application of IoT has a significant impact on understanding the tourist experience, since it can reshape travelling experiences throughout the entire trip; they conclude with directions for both academics and practitioners in several tourism fields, such as destination management, hospitality, transportation and intermediaries.

The second chapter is the work of Ilma Aulia Zaim and studies a specific market segment, young British tourists aged 18–35, and how these pursue tourism-related information sources that eventually affect their travel decision-making prior to visiting a destination. The study offers significant insight into DMOs regarding young British travellers' characteristics and offers suggestions to them such as the types of information they should share on their websites and which of them are really valuable for the particular tourism market segment.

The third chapter, by Bernardo Borges and Rui Augusto Costa, deals with the DMOs and the tourism marketing channels in the Centre of Portugal. The chapter assesses the significance of the digital marketing strategy of the DMOs responsible for the Centre of Portugal and provides insights regarding how this strategy can affect tourists' information search during the pre-trip as well as at the destination. The study concludes with useful information for DMOs regarding the main promotion channels that they should use for attracting tourists to the region.

The fourth chapter is the work of Francesco Calza, Michele Simoni, Annarita Sorrentino and Mariapina Trunfio, who study the role of participatory planning in repositioning a southern Italy destination, Centola-Palinuro in the Campania region. This contribution deals with a demand-driven destination that suffers from several challenges such as seaside seasonality, mass-market demand and low spending. The case argues that a participatory planning process that involves local stakeholders can have a significant impact in the repositioning of the destination by reshaping the tourism offering in the area through stakeholder engagement.

The fifth chapter, by Jesús Barreal Pernas and Gil Jannes, deals with gastro-tourism in Spain and studies the international tourist profile in the Spanish wine and olive oil Protected Destinations of Origin (PDOs). Specifically, the study focuses on regions that have such PDOs and examines the differences and similarities among the profiles of international tourists that visit them. The study concludes with a classification of four clusters according to demographic, travel and motivational characteristics of tourists in the particular area.

The sixth chapter deals with the case of the Alps. Thomas Bausch discusses the hybridity of tourist destinations from a supply-and-demand perspective, focusing on the region of European Alps as a representative mountainous destination in Europe. The chapter suggests that the Alps is a multi-faceted destination area, shared by various countries of western Europe, which offers a hybrid tourism product in different seasons. The author argues that the region is characterized by three types of destination hybridity: uniform non-hybrid, binary-hybrid and multiple-hybrid, and concludes with suggestions regarding managing and marketing these types of destinations.

The seventh chapter is by Nikolaos Boukas and studies the introduction of new products in traditional mass destinations by emphasizing rural tourism on the Greek mountainous destination of Kalavrita. Despite the important role of tourism for a country like Greece, the majority of the attention by tourism marketing has focused mainly on its islands and leaves the mainland areas unexplored in terms of tourism. The chapter suggests that rural tourism could provide great opportunities to areas such as mountainous ones and concludes with recommendations for DMOs and destination marketing professionals with regard to tourism product development.

Chapter 8 deals with two western European countries, France and Italy. In this chapter, Jacey Choe studies US tourists' food neophobia and their local food experiences in France and Italy. This study explores whether tourists have different levels of food neophobia based on their socio-demographic profile and investigates characteristics of different food neophobia groups, as well as their behavioural profile. The chapter provides important insight for French and Italian destinations, regarding US tourists' local food perception experiences.

Chapter 9, by Elitza Iordanova, investigates the destination image of Linz, Austria, regarding its Nazi dark past. According to this study, the attempts of local authorities and destination marketers to distance the dark past of the town have helped to establish an image for Linz as a contemporary art and cultural destination for international visitors; nonetheless, Austrian visitors still have some level of prejudice regarding the heritage of the city. The study concludes with implications for marketers, tourism professionals and local authorities of places with a dark past.

Chapter 10, by Carina Ren and Elizabeth Cooper, investigates the rhythms of connectivity and user-generated content in the marketing of Greenland, a self-ruled Arctic nation under the Danish Commonwealth. The study investigates how users negotiate online presence and social media content while travelling to Greenland, based on professional social media influencers. The study concludes with implications for marketing less-connected European destinations such as Greenland, as well as destinations characterized by partnerships between DMOs and social media influencers.

Finally, Chapter 11, by Jordi Arcos-Pumarola and Marta Conill-Tetuà, deals with the promotion of intangible cultural heritage through social networks and investigates the case of the Fête de l'Ours festival that is held in the Vallespir region of France. This study recognizes the elements highlighted by locals, visitors, professionals and official bodies in their Instagram accounts and investigates the way users share and explain their experiences of the festival. It concludes with strategies for destination marketers regarding the promotion of intangible heritage and its meanings via the context of social media.

Overall, the volume, through the demonstration and discussion of various marketing issues and paradigms in a number of diverse destinations in western Europe, hopes to enrich existing literature in tourism by contributing to further understanding and knowledge regarding destination marketing matters, trends, practices and challenges. The methodologies, implications and conclusions used in each presented study could inspire tourism students, academics and practitioners regarding how effective marketing evolves and adds value to the overall management of destinations such as those in the western European region.

References

Costa, C., Panyik, E. and Buhalis, D. (2013) *Trends in European Tourism Planning and Organisation*. Vol. 60. Channel View Publications, Clevedon, UK.

e Silva, F.B., Herrera, M.A.M., Rosina, K., Barranco, R.R. and Freire, S. *et al.* (2018) Analysing spatiotemporal patterns of tourism in Europe at high-resolution with conventional and big data sources. *Tourism Management* 68(3), 101–115. DOI: 10.1016/j.tourman.2018.02.020.

Higgins-Desbiolles, F. (2020) Socialising tourism for social and ecological justice after COVID-19. *Tourism Geographies* 22(3), 610–623. DOI: 10.1080/14616688.2020.1757748.

John, P. (2001) *Local Governance in Western Europe*. Sage, London.

Kuščer, K. and Mihalič, T. (2019) Residents' attitudes towards overtourism from the perspective of tourism impacts and cooperation: the case of Ljubljana. *Sustainability* 11(6), 1–16. DOI: 10.3390/su11061823.

Oriade, A. and Schofield, P. (2019) An examination of the role of service quality and perceived value in visitor attraction experience. *Journal of Destination Marketing & Management* 11(3), 1–9. DOI: 10.1016/j.jdmm.2018.10.002.

Peeters, P., Gossling, S., Klijs, J. and Milano, C. (2021) *Research for TRAN Committee-Overtourism: Impact and Possible Policy Responses*. European Parliament, Policy Department for Structural and Cohesion Policies, Brussels.

Ritchie, J.B. and Crouch, G.I. (2003) *The Competitive Destination: A Sustainable Tourism Perspective*. CAB International, Wallingford, UK.

UNWTO (2020) *International Tourism Highlights*, 2020 Edition. World Tourism Organization, Madrid.

WTTC (2020) *Country Data*. World Travel & Tourism Council.

1 How Does Internet of Things (IoT) Affect Travel Experience?

Mengyun Hu*, Eleonora Pantano and Nikolaos Stylos

University of Bristol, Bristol, UK

1.1 Introduction

Tourism, as a service-intensive industry, has always been following closely and implementing state-of-the-art technologies. Nowadays, this sector tends to integrate the latest information technology applications such as Internet of Things (IoT) (Benckendorff *et al.*, 2019). In fact, the entire tourism and hospitality industry has been transformed by the increasing implementation of technologies. Neuhofer *et al.* (2015) indicate that innovative technologies have not only become an integral part of tourism but have also fundamentally reshaped the way travel is planned, business is conducted and tourism services are provided and received. Moreover, previous studies have found that traveller behaviours were changed due to the development of new technologies, which also result in new types of travel experiences (Jeong and Shin, 2019). For instance, the growing capabilities of cloud computing and IoT have revolutionized travellers' experiences by bringing together information searches, interaction, social networking and remote-control functionalities to travellers (Pesonen and Horster, 2012; Buhalis and Foerste, 2015; Huang *et al.*, 2017; Buhalis *et al.*, 2019; Jeong and Shin, 2019).

In recent years, the impact of IoT on travellers' experience has been widely recognized by researchers and practitioners from various industries (Gretzel *et al.*, 2015; Balaji and Roy, 2017; Novak and Hoffman, 2018). IoT refers to numerous objects that can be equipped with sensors, digital logic and processing technologies to identify and integrate into communication networks to achieve common goals (Asghari *et al.*, 2019). In other words, it allows objects to act smartly and make informed decisions under many circumstances (Salman and Jain, 2019). In the tourism industry, IoT has created new ways of interaction, data collection and analysis, and thus new opportunities for value creation and management (Xiang *et al.*, 2015a). This leads to new levels of competitiveness among service providers in this industry and transforms the traveller's

*mengyun.hu@bristol.ac.uk

DOI: 10.1079/9781789248753.0001

experience through providing high-quality services. In other words, it implies that IoT can trigger limitless possibilities for the development of the tourism industry by enhancing travellers' experience (Gretzel *et al.*, 2015).

Specifically, tourism destinations provide IoT ecosystems with mobile applications and sensors to support travellers in interacting with their surroundings (e.g. information search and navigation, interpretation of cultural attractions etc.) (Tussyadiah *et al.*, 2018). Physical limits imposed now have a limited influence on tourist experience at the destination (Tan, 2017). Furthermore, the coronavirus pandemic has placed great limits on physical contacts and face-to-face interaction, thereby giving birth to travellers' Covid-19-related safety concerns. Many destinations have been facilitating the use of IoT-enabled applications (e.g. Contact Tracing App (CTA)) as one form of safety and security strategy, to improve travellers' experiences via eliminating their safety concerns (Pillmayer *et al.*, 2021). Thus, it is possible to argue that the advances in IoT technologies give travellers a novel way of enjoying trips at a specific destination. Along these lines, it becomes imperative to understand the extent, the triggering factors and the influences of IoT usage on the tourism industry.

The purpose of this chapter is to illustrate an in-depth understanding of the use of IoT in tourism and its impact on the travel experience. In the next section, we first review the previous research on the IoT application in tourism-related sectors, including destination management, hospitality, transportation and intermediaries. Subsequently, the chapter critically discusses the relationship between IoT and travel experience. Next, we provide a future agenda for IoT applications in the tourism sector, suggesting the solutions and recommendations for the use of IoT in tourism. The final section draws conclusions and makes final key points.

1.2 IoT Applications in Tourism

1.2.1 IoT applications in destination management

Destination management can be seen as a vital process that brings together different stakeholders and authorities to work towards a common goal to ensure the viability and integrity of destinations (Benckendorff *et al.*, 2019). Digital technologies such as IoT can be useful in supporting destination management organizations (DMOs) to connect travel markets and travellers, as well as understanding today's travellers' preferences and behaviours (Jeong and Shin, 2019). Thus, in recent years an increasing number of destinations have adopted IoT technologies as a strategic tool for destination management to support tourism experience creation and enhance efficiency and effectiveness (Almobaideen *et al.*, 2017; Li *et al.*, 2017; Cimbaljević *et al.*, 2019).

One of the common IoT applications in destination management is attractions. For example, Disney developed and released an exclusive product using IoT technologies in 2014, named MagicBand. Using MagicBand, visitors were able to access many facilities, including the theme park and their hotel room, with just a touch of the wearable wristband, as well as buying food and merchandise (Borkowski *et al.*, 2016). Moreover, in San Francisco's De Young Museum, visitors can access audio-visual materials and testimonials revealing the hidden story behind Keith Haring's artworks by using Google Glass (Tussyadiah *et al.*, 2018). Moreover, Chianese and Piccialli (2014) propose an IoT-based smart museum system with user-friendly intelligent apps to assist visitors. This system enhances visitors' experience via provision of the relevant, real-time and precise information with respect to the particular segment and exhibition. Literature suggests that such may be the case with other attractions and entertainment as well (Alletto *et al.*, 2015; Wang *et al.*, 2016; Spachos and Plataniotis, 2020).

In addition, IoT has been used for data analysis in the process of destination management. With the massive number of devices connected to the internet, a huge amount of real-time and context-aware data can be automatically produced and analysed (Xu *et al.*, 2014). In the process of destination management, IoT applications are expected to automatically collect and analyse the data about the destinations and travellers, which may help destinations become better and more effectively managed (Ammirato *et al.*, 2018). In addition, DMOs can utilize these data to adjust their operation and strategies accordingly to enhance the travel experience and their competitiveness (Buhalis and Amaranggana, 2015). IoT technologies also enable easy access to and interaction with context-aware information such as transportation, attractions, tours and hotels. Thus, they facilitate travellers' interaction with and integration into their surroundings, increasing the quality of experience at the destination (Benckendorff *et al.*, 2019).

1.2.2 IoT applications in hospitality

As the goal of hospitality is to ensure a higher level of customer satisfaction via highly personalized offerings (Berezan *et al.*, 2015), IoT applications play an essential role in the hospitality industry. More specifically, it helps service providers to, on the one hand, improve operational efficiency and effectiveness with lower costs (Wu and Cheng, 2018) and, on the other, ensure guest satisfaction by providing a more personalized guest experience and customized stay (Amer and Alqhtani, 2019). For example, Yuksel *et al.* (2020) proposed an IoT-based Paperless Buffet Management system for restaurants which is context-aware, customer-friendly and staff-oriented. This system reveals that IoT can allow restaurants to forecast the crowd, update the menu and meal information easily, and thus improve operational efficiency as well as service quality.

To enhance the guest experience, several exciting IoT solutions have been applied in hotels over the last decade, such as self check-in and check-out, keyless entry and room-controlling, used in Hilton and Starwood for example (Buhalis and Leung, 2018; Car *et al.*, 2019). Moreover, Peninsula Hotels have applied IoT technologies to provide integrated services like tablet-controlled devices. These allow the guest to adjust room settings based on their preferences, order room service, arrange transportation and find nearby attractions through in-room tablets (Kansakar *et al.*, 2019).

1.2.3 IoT applications in transportation

Transportation is an integral part of the tourism industry (Marlina and Natalia, 2017). To date, IoT technologies have been applied to various aspects of the transportation industry which significantly affect the tourism industry, such as intelligent transportation systems, shared mobility and smart airports (Weng *et al.*, 2017; Lykou *et al.*, 2019; Benckendorff *et al.*, 2019).

An intelligent transportation system integrated with IoT technologies during recent years has offered a faster, safer, more efficient and more user-friendly transport for travellers (Benckendorff *et al.*, 2019). More specifically, with sensors embedded in such an intelligent transport system, physical objects can be monitored in real time, and enormous data can be generated (Herrera-Quintero *et al.*, 2018). For example, a vast amount of traffic data can be collected from physical devices, such as GPS, traffic lights, induction coils and video camera. This can assist in the improvement of public transport service quality and contribute to higher tourist satisfaction (Patel *et al.*, 2019).

Shared mobility service has experienced significant growth in the transportation industry by using IoT technologies over the last decade (Clewlow and Mishra, 2017). As part of the tourism product, this allows tourists to use more diverse forms of transportations than ever before, such as bike-sharing, scooter-sharing and car-sharing (Stocker and Shaheen, 2018). In addition, many taxi service companies, such as Uber and Lyft, have adopted IoT technologies to develop user-centric mobile applications, which combine real-time and context-aware services with a payment system as well as a review platform (Benckendorff *et al.*, 2019). They also allows tourists to order taxis on their smartphone at any time (Weng *et al.*, 2017), trace the vehicle and receive a notification when it arrives (Benckendorff *et al.*, 2019). This, in turn, offers tourists more convenient, comfortable and enjoyable travel experiences (Meng *et al.*, 2020).

Smart airport is becoming a buzzword in the era of IoT. It can be seen as a 'connected airport' which is important for airport operations and passenger experience (Lykou *et al.*, 2019). For instance, many airports such as Hong Kong International Airport have developed IoT-based mobile applications that provide visitors with many functionalities, including updates on gates, flight times, baggage carousel numbers and personal recommendations about nearby restaurants and shops (Hemdi *et al.*, 2016). Besides, with the use of IoT, self-service systems in airports have been developed that provide self-service capabilities for a wide range of actions such as self and remote check-in (Hemdi *et al.*, 2016), self and remote bag drop (Mallick *et al.*, 2018), self-boarding and e-Border pass (Mayer, 2019). These smart applications in airports can reduce passengers' wait time, assist operation teams in tracking passengers, and appropriately arrange resources to ensure passenger activities run smoothly (Luis *et al.*, 2018).

1.2.4 IoT applications in travel intermediaries

Travel intermediaries are firms that connect travel suppliers with customers, who are considered as intensive users of information (Benckendorff *et al.*, 2019). In the era of IoT, the tourism industry structure has been transformed due to its fragmented and information-intensive nature (Sharma *et al.*, 2020). In particular, the spread of IoT technologies has transformed the consuming styles and distribution channels in the tourism sector. Travel service providers need to use smart technologies such as mobile applications, websites, social media and Augmented Reality (AR) for attracting new customers (Mohseni *et al.*, 2018), and enhance the quality of service as well as operational efficiency (Sharma *et al.*, 2020). Besides, the application of IoT in intermediaries enables the collection of data on tourist experiences and preferences. It enables travel intermediaries to develop more suitable tourism spot packages and improve tourist loyalty and engagement (Car *et al.*, 2019). For example, many travel agencies and tour operators have developed IoT-based mobile applications such as Ctrip and Qunar that allow travellers to find, compare and book tickets, hotel rooms and rental cars and purchase other travel products. These provide real-time and convenient services for travellers as well as gathering associated information about their preferences and behavior.

1.3 IoT and Travel Experience

1.3.1 The role of IoT in the travel experience

Prior studies point out that travel can be seen as a 'linear' process, which includes three phases: pre-trip; during-trip; and post-trip (Wang and Fesenmaier, 2013; Neuhofer *et al.*, 2015; Kheiri and Nasihatkon, 2015). Travel is a series of activities linked to these three phases, including information searching, travel planning, reservations making, visiting, dining and reflection (Huang *et al.*, 2017). Hence, from the temporal viewpoint, the travel experience is often referred to as an 'activity-based' process (Neuhofer *et al.*, 2012).

Advances in IoT technologies have brought a wide range of smart devices and intelligent services into the tourism industry (Buhalis and Amaranggana, 2015; Gretzel *et al.*, 2015; Buhalis, 2019). It affects travel activities within all stages of the travel process. During the preparation phase of a trip, IoT provides travellers with personalized information, giving them the ability to make wise and informed travel choices such as deciding which method of travel to take. It also ensures an effective transaction-oriented process such as reservation-making and payment-taking and provides a cross-platform for travellers to communicate with service providers (Huang *et al.*, 2017). During the travelling period, IoT provides travellers with rich, real-time, and context-aware information due to its higher level of connectivity and accessibility. It keeps travellers informed of their current situation and assists them in making more effective decisions (Verma and Shukla, 2019). At the post-trip stage, IoT applications allow travellers to share their experiences with other people and service providers more easily.

In an effort to better understand the impact of IoT on tourist demands, Benckendorff *et al.* (2019) proposed a Mobile Technology Ecosystem which incorporates some key IoT-enabling technologies such as Radio-Frequency Identification (RFID), Near-Field Communication (NFC), Global Positioning Systems (GPS) and some sensing technologies. This system has demonstrated that IoT technologies can help change tourists' behaviour and demands by addressing a wide range of functional and informational needs. In this system, travellers use their mobile devices for a variety of management functions, including time management, navigation, dining reservations, accommodation bookings, flight check-in and itinerary management. By taking advantage of IoT, tourists could simply use their mobile devices to receive rich information, location-based, customized and interactive services (Buhalis and Amaranggana, 2015). The combination of everyday objects transmitting data and people carrying sensing devices allow for the co-creation of travel experiences to occur in real time. Thus, it is undoubted that the use of IoT technologies in travel can significantly affect the tourist experience by generating rich and real-time intelligence about tourists' needs and wants and responding to them.

Despite insufficient academic studies having specifically investigated the influence of IoT on the travel experience, some scholars have demonstrated that smart technologies (e.g. geo-based technology, NFC, RFID, smartphones, wearable devices) have significantly affected tourists' experience and satisfaction (Tussyadiah and Zach, 2012; Wang *et al.*, 2016; Tussyadiah *et al.*, 2018). Gretzel *et al.* (2015) point out that IoT is considered as one of the fundamental smart tourism technologies, which is reshaping the way tourism services are provided and received. Almobaideen *et al.* (2017) further indicate that IoT offers a set of innovative services which make travellers' trips more efficient and safer, such as location-based suggestions including restaurants, souvenir shops, shopping malls and banks; and even a restroom can be provided during the trip. Thus, Li *et al.* (2017) have explained that IoT technologies play an essential role in reshaping tourist experience because they offer travellers a more personalized and ubiquitous tourism service that might exceed their expectations.

Based on these arguments, it is reasonable to conclude that the role of IoT in the travel experience is significant due to its unique abilities, such as informing, which refers to the ability to source and access information while travelling; contextualizing, which refers to the ability to offer contextual information based on contextual sensors; personalizing, which refers to the ability to combine contextual information with user data to provide personalized service; remote controlling, which refers to the ability to control something from distance; and communicating, which refers to communication via social media or messaging. As shown in Table 1.1, IoT-based travel applications can affect tourists' travel activities throughout the entire trip.

Table 1.1. The role of IoT in the travel experience.

IoT capabilities	Description	Travel applications/examples	Travel phases
Informing	The ability to source and access information while travelling	• Attraction and destination guides • Navigation	During-trip
Contextualizing	The ability to offer contextual information based on the contextual sensors	• Push real-time notifications relative to location and context • Live travel information (e.g. flight status, weather, safety, events and offers) • Wearable devices (applied in the museum, theme park, galleries, etc.)	During-trip
Personalizing	The ability to combine contextual information with user data to provide personalized service	• Recommender system (hotel, travel packages, flight tickets, etc.) • Location-based suggestions • Personalized hotel rooms	Pre-trip, during-trip
Remote controlling	The ability to control something from distance	• Smart hotel (self check-in/out, keyless entry, control the room from a mobile phone)	During-trip
Communicating	The ability to communicate, such as social media and messaging	• Mobile social media (WeChat, Facebook, Tripadvisor, etc.)	Pre-trip, during-trip, post-trip

1.3.2 IoT and perceived value

Perceived value has been considered as one of the most important factors affecting consumers' post-purchase responses and behaviours (Hong *et al.*, 2017; Pham *et al.*, 2018), as well as satisfaction and loyalty (Ashraf *et al.*, 2018; Chen and Lin, 2019; El-Adly, 2019). In the tourism literature, perceived experience value has also received increasing attention from researchers (Kim and Thapa, 2018; Kim *et al.*, 2019a; Oriade and Schofield, 2019; Loureiro *et al.*, 2020). Kim *et al.* (2015) have explained that perceived value of travel experience (PVTE) can be seen as the process by which a visitor receives, selects, arranges and interprets information based on their various experiences while travelling, to create a meaningful picture of the value of travel experiences.

Moreover, Pandža Bajs (2015) illustrates that perceived value is a concept that offers insight into how consumers perceive given products or services and provides guidance as to how to create the product or service in order to nalyzi customers' needs and expectations. Thus, understanding travellers' perceived value of travel experience with IoT is expected to provide insights into travellers' post-purchase responses and behaviours in the era of digitalization.

This study develops the perceived value into two dimensions: functional and emotional value. Functional value refers to the rational and economic valuations made by individual travellers (Sánchez-Fernández and Iniesta-Bonillo, 2007). In the context of IoT usage, this also means the functional and instrumental benefits delivered by using IoT technologies while travelling, focusing on individual tourism consumers' perception of utility and performance (Jamal *et al.*, 2011). Emotional value has been defined as a product's or service's ability to arouse feelings or affect states (Sheth *et al.*, 1991). Following this concept, this study defines emotional value as travellers' perceived specific feelings with the use of IoT technologies while travelling, such as comfort, security, pleasure, excitement and enjoyment (Jamal *et al.*, 2011).

For instance, an increasing number of IoT devices in the tourism industry create opportunities for generating considerable amounts of data, known as big data, regarding travellers' needs and preferences (Stylos and Zwiegelaar, 2019; Stylos *et al.*, 2021). Tourism stakeholders can get valuable insights into travellers' preferences, ideas, values and behaviours from analysing these data, and thus provide the service in a more efficient and effective way (Kim *et al.*, 2019b; Xiang *et al.*, 2015b). This, in turn, leads to a higher level of perceived value of travel experience by tourists.

1.4 A Future Agenda for IoT Applications in the Tourism Sector

Although much has been written about IoT in previous studies, there are still many IoT applications that are far from being fully addressed in the existing tourism literature. A discussion of the future agenda for IoT in tourism is therefore useful for both practitioners and scholars. Given this background, this chapter considers how IoT might change the future of tourism in four tourism-related sectors, including (i) destination management; (ii) hospitality (iii) transportation; and (iv) intermediaries.

1.4.1 Destinations management

In recent years, IoT technologies are shifting destination management systems to smart destinations because of the connectivity of devices and systems that travellers can, in turn, customize (Buhalis *et al.*, 2019). It can further be posited that the application of IoT is reshaping the future of destinations by revolutionizing the way of interactions among travellers, services, products and destinations.

Based on the roles of IoT applications in tourism and travel experiences, it becomes imperative for researchers in the future to understand the extent, triggering factors and influences of IoT usage in the tourism industry. In addition, although the adoption of IoT technologies will be beneficial for the destinations and travellers, further research needs to ascertain the negative impacts of IoT on destination management, to help destination managers to understand how to facilitate IoT adoptions and apply them properly.

Li *et al.* (2017) point out that one of the most prominent future challenges to destinations is adapting business strategies, tactics and initiatives to the advancement of new technologies to meet consumer expectations. Travellers' behaviour, preferences and expectations are changing with the use of technology. Thus, destination service providers should pay more attention to understanding the needs, wishes and desires of travellers in the era of IoT, and create a successful future through adapting marketing strategies with IoT. In addition, DMOs also need to future-proof their technology framework so that the destination management system can be easily upgraded in tandem with the changing IoT technologies.

1.4.2 Hospitality

Kansakar *et al.* (2019) point out that the future of the hospitality industry is being changed by the current advances in IoT technology. Many smart hospitality services (e.g. smart showers) can be expected to come about as IoT technologies are becoming more mature (Kansakar *et al.*, 2019). Understanding how IoT shapes the future of the hospitality industry provides valuable insights into business innovation (Haaker *et al.*, 2021).

Some previous studies predict that potential future hospitality services should include the personalization of experiences and digitalization of services (Neuhofer *et al.*, 2015; Kansakar *et al.*, 2019; Buhalis *et al.*, 2019). Thus, future scholars could empirically investigate the relationship between IoT applications and guest experiences. This would provide insights into travellers' attitudes and responses to IoT usage. In this regard, researchers would be able to suggest improvements for designing IoT solutions for the hospitality industry.

In line with Pestek and Sarvan (2020), the hospitality industry should be urged to be more future-oriented and prepare fully for the planning of technology adaption. For example, in the post-pandemic era, safety will be a major concern for most travellers. Hoteliers should invest more in IoT-enabled, contactless technologies which can help increase guests' comfort level and confidence in a safer stay. In addition, security and privacy issues are major concerns among the difficulties besetting the IoT industry, especially at the consumer level. Guests appreciate the prospective benefits but are concerned with privacy and usability (Car *et al.*, 2019). Thus, guaranteeing the privacy and security of guest information is essential for the hospitality industry to improve customer satisfaction and build trust with guests.

1.4.3 Transportation

The application of IoT is predicted to have a transformative effect on transportation and create seemingly endless possibilities and benefits for future travellers (Sodhro *et al.*, 2019). For example, some uncertainties around travel can be removed through the use of IoT. In particular, smart devices can offer visitors up-to-date information about transport or congestion at resorts.

According to Benckendorff *et al.* (2019), transport not only refers to the utility of moving from place to place, but it can also be integral to the tourism experience. Thus, future studies on the influence of IoT on the transportation industry should consider a deeper engagement with travel experiences. In addition, Shaheen and Cohen (2018) state that realizing the full potential of technology on transportation requires us to consider how users' travel behaviour will respond to it. Providing a deeper understanding of the link between the IoT innovation process and travellers' behaviour is necessary for future research.

In the transportation industry, IoT can be seen as the cornerstone of future sustainable transport systems (Chavhan *et al.*, 2019). Nowadays, IoT technologies are changing in a significant and unforeseen way. Thus, similar to destination management, the technology framework of transportations needs to be future-proofed so that the destination management system can be updated quickly with the changing IoT landscape. Furthermore, understanding how travellers make decisions and interact with IoT technologies would be helpful for the transportation industry to meet the traveller's future needs most optimally.

1.4.4 Intermediaries

Technological innovation unites all relevant stakeholder groups in the tourism service ecosystem, which has witnessed a vertiginous transformation of the intermediaries' industry (Stylos, 2019; Romero and Tejada, 2020). However, tourism literature on the relationship between IoT technology and intermediaries remains limited. Future scholars could further investigate how IoT affects intermediaries, to provide insights into the potential impact of IoT on the business model, business opportunities and future of intermediaries in the tourism industry. Indeed, the role of IoT in supporting travel during Covid-19 should be examined. It would offer insights into how intermediaries use IoT to innovate their service offerings.

In a post-Covid world, travel will become increasingly complex; future travellers might have higher expectations and safety concerns for their travel. Technological innovation will play an essential role in rebuilding traveller

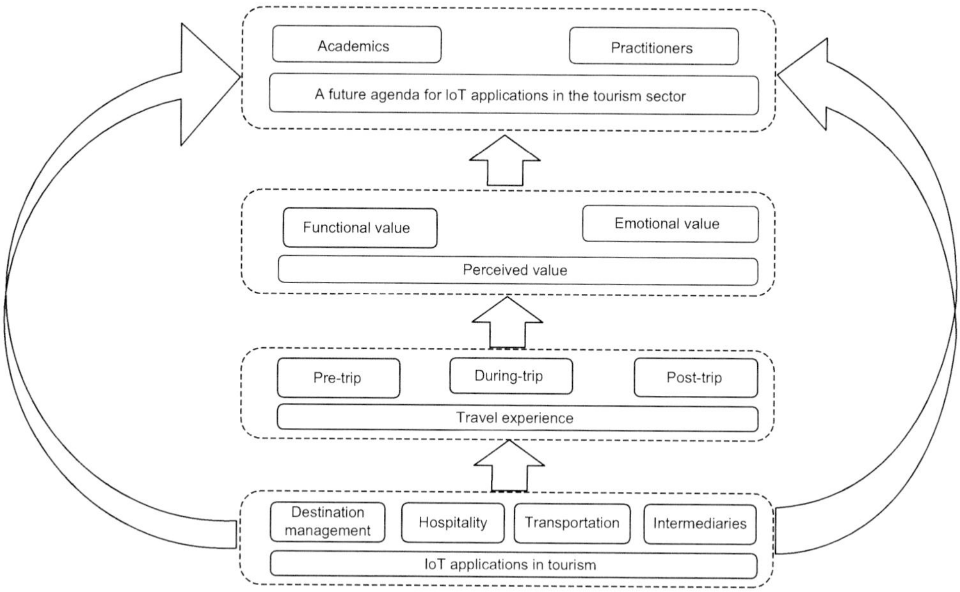

Fig. 1.1. IoT applications and travel experience.

confidence. Thus, future intermediaries should invest more in IoT applications to meet travellers' requirements in the post-Covid era. Furthermore, intermediaries should pay more attention to finding out how IoT technologies are creating new habits and behaviours for travellers and then adjusting their business strategies appropriately.

1.5 Conclusion

This chapter provides valuable insights into how IoT technologies can reshape travelling experiences throughout an entire trip. As shown in Fig. 1.1, we develop a framework that supports a better understanding of the relationship among IoT applications, travel experience and perceived value, and future agendas. In particular, a number of IoT solutions have been developed in various industries linked to tourism, such as destination management, hospitality, transportation and intermediaries that have changed the way in which services are provided and received in the tourism industry (e.g., Apanasevic and Markendahl, 2018; Benckendorff *et al.*, 2019; Car *et al.*, 2019). Given this background, it is reasonable to conclude that IoT plays a significant role in travel experiences, which in turn affect travellers' perceived value. In this vein, we set a future research agenda for tourism scholars regarding IoT, as well as suggest an improvement for IoT applications in the tourism sector.

Over the last decade, the value of IoT in the tourism industry has been widely acknowledged by both academics and practitioners. However,

Table 1.2. A future agenda for IoT applications in the tourism sector.

Tourism sectors	Academics	Practitioners
Destination management	• To understand the extent, the triggering factors and influences of IoT usage on destination management • To ascertain the negative impact of IoT on destination management	• To understand exactly what tourists need and want in the era of IoT • To adjust business strategies to the development of IoT
Hospitality	• To examine the role of IoT in guest experience from the traveller's perspective • To suggest an improvement for the way to design IoT solutions for the hospitality industry	• To invest more in IoT- enabled, contactless technologies to reduce the impact of Covid-19 • To guarantee the privacy and security of guest information
Transportation	• To consider a deeper engagement on travel experience • To provide a deeper understanding of the link between the IoT innovation process and travellers' behaviour	• To future-proof the technology framework of transportations • To understand how travellers make decisions and interact with IoT technologies
Intermediaries	• To understand the potential impact of IoT on the business model, business opportunities and future of intermediaries • To understand the role of IoT in supporting travel during the pandemic	• To design IoT solutions to meet travellers' needs in a post-Covid-19 world • To find out how IoT is creating new habits and behaviours for travellers

the tourism literature on the relationship between IoT technologies and travel experiences in different tourism contexts remains limited. Hence, this chapter sets up a future agenda which identifies several promising directions for future studies on IoT applications in the different tourism sectors, including destination management, hospitality, transportation and intermediaries, as summarized in Table 1.2. This chapter has also provided a couple of recommendations for tourism managers to enhance travellers' experience and improve their competitiveness. For example, we suggest hospitality managers invest more in IoT-enabled, contactless technologies to reduce the impact of Covid-19, which might in turn improve travellers' satisfaction. Thus, tourism managers could learn from this study and strengthen the basis of their practice. Furthermore, this study takes a summarizing perspective and therefore empirical analysis is lacking. This in turn may encourage more studies that can address critical cases mentioned above.

References

Alletto, S., Cucchiara, R., Del Fiore, G., Mainetti, L., Mighali, V. *et al.* (2015) An indoor location-aware system for an IoT-based smart museum. *IEEE Internet of Things Journal* 3, 244–253.

Almobaideen, W., Krayshan, R., Allan, M. and Saadeh, M. (2017) Internet of Things: geographical routing based on healthcare centers' vicinity for mobile smart tourism destination. *Technological Forecasting and Social Change* 123, 342–350.

Amer, M. and Alqhtani, A. (2019) IoT applications in smart hotels. *International Journal of Internet of Things and Web Services* 6.

Ammirato, S., Felicetti, A.M., Della Gala, M., Raso, C. and Cozza, M. (2018) Smart tourism destinations: Can the destination management organizations exploit benefits of the ICTs? Evidences from a multiple case study. In: CamarinhaMatos, L., Afsarmanesh, H. and Rezgui, Y. (eds) *Collaborative Networks of Cognitive Systems. PRO-VE 2018. IFIP Advances in Information Communication Technology*. Springer, Cham, Switzerland, pp. 623–634.

Apanasevic, T. and Markendahl, J. (2018) The value of mobile ticketing from a public transport perspective. *Journal of Payments Strategy & Systems* 11, 292–305.

Asghari, P., Rahmani, A.M. and Javadi, H.H.S. (2019) Internet of Things applications: a systematic review. *Computer Networks* 148, 241–261.

Ashraf, S., Ilyas, R., Imtiaz, M. and Ahmad, S. (2018) Impact of service quality, corporate image, and perceived value on brand loyalty with presence and absence of customer satisfaction: a study of four service sectors of Pakistan. *International Journal of Academic Research in Business and Social Sciences* 8, 452–474.

Balaji, M.S. and Roy, S.K. (2017) Value co-creation with Internet of Things technology in the retail industry. *Journal of Marketing Management* 33, 7–31.

Benckendorff, P., Xiang, Z. and Sheldon, P. (2019) *Tourism Information Technology*. CAB International, Boston, Massachusetts.

Berezan, O., Raab, C., Tanford, S. and Kim, Y.S. (2015) Evaluating loyalty constructs among hotel reward program members using eWOM. *Journal of Hospitality & Tourism Research* 39, 198–224.

Borkowski, S., Sandrick, C., Wagila, K., Goller, C., Ye, C. *et al.* (2016) Magicbands in the magic kingdom: customer-centric information technology implementation at Disney. *Journal of the International Academy for Case Studies* 22, 143.

Buhalis, D. (2019) Technology in tourism from information communication technologies to eTourism and smart tourism towards ambient intelligence tourism: a perspective article. *Tourism Review* 75, 267–272.

Buhalis, D. and Amaranggana, A. (2015) Smart tourism destinations enhancing tourism experience through personalisation of services. In: Tussyadiah, I. and Inversini, A. (eds) *Information and Communication Technologies in Tourism 2015*. Springer, Cham, Switzerland, pp. 377–389.

Buhalis, D. and Foerste, M. (2015) SoCoMo marketing for travel and tourism: empowering co-creation of value. *Journal of Destination Marketing & Management* 4, 151–161.

Buhalis, D. and Leung, R. (2018) Smart hospitality: interconnectivity and interoperability towards an ecosystem. *International Journal of Hospitality Management* 71, 41–50.

Buhalis, D., Harwood, T., Bogicevic, V., Viglia, G., Beldona, S. *et al.* (2019) Technological disruptions in services: lessons from tourism and hospitality. *Journal of Service Management* 30, 484–506.

Car, T., Stifanich, L.P. and Šimunić, M. (2019) Internet of Things (IoT) in tourism and hospitality: opportunities and challenges. *Tourism in South East Europe* 5, 163–175.

Chavhan, S., Gupta, D., Chandana, B.N., Khanna, A. and Rodrigues, J.J. (2019) IoT-based context-aware intelligent public transport system in a metropolitan area. *IEEE Internet of Things Journal* 7, 6023–6034.
Chen, S.C. and Lin, C.P. (2019) Understanding the effect of social media marketing activities: the mediation of social identification, perceived value, and satisfaction. *Technological Forecasting and Social Change* 140, 22–32.
Chianese, A. and Piccialli, F. (2014) Designing a smart museum: when cultural heritage joins IoT. *In: 2014 Eighth International Conference on Next Generation Mobile Apps, Services and Technologies*, IEEE, pp. 300–306.
Cimbaljević, M., Stankov, U. and Pavluković, V. (2019) Going beyond the traditional destination competitiveness: reflections on a smart destination in the current research. *Current Issues in Tourism* 22, 2472–2477.
Clewlow, R. and Mishra, G.S. (2017) Shared mobility: current adoption, use, and potential impacts on travel behavior. *In: Transportation Research Board 96th Annual Meeting*, Washington, DC.
El-Adly, M.I. (2019) Modelling the relationship between hotel perceived value, customer satisfaction, and customer loyalty. *Journal of Retailing and Consumer Services* 50, 322–332.
Gretzel, U., Sigala, M., Xiang, Z. and Koo, C. (2015) Smart tourism: foundations and developments. *Electronic Markets* 25, 179–188.
Haaker, T., Ly, P.T.M., Nguyen-Thanh, N. and Nguyen, H.T.H. (2021) Business model innovation through the application of the Internet-of-Things: a comparative analysis. *Journal of Business Research* 126, 126–136.
Hemdi, M.A., Rahman, S.A.S., Hanafiah, M.H. and Adanan, A. (2016) Airport self-service check-in: the influence of technology readiness on customer satisfaction. *In: Proceedings of the 3rd International Hospitality and Tourism Conference & 2nd International Seminar on Tourism*, Bandung, Indonesia, pp. 10–12.
Herrera-Quintero, L.F., Vega-Alfonso, J.C., Banse, K.B.A. and Zambrano, E.C. (2018) Smart ITS sensor for the transportation planning based on IoT approaches using serverless and microservices architecture. *IEEE Intelligent Transportation Systems Magazine* 10, 17–27.
Hong, J.C., Lin, P.H. and Hsieh, P.C. (2017) The effect of consumer innovativeness on perceived value and continuance intention to use Smartwatch. *Computers in Human Behavior* 67, 264–272.
Huang, C.D., Goo, J., Nam, K. and Yoo, C.W. (2017) Smart tourism technologies in travel planning: the role of exploration and exploitation. *Information & Management* 54, 757–770.
Jamal, S.A., Othman, N.A. and Muhammad, N.M.N. (2011) Tourist perceived value in a community-based homestay visit: an investigation into the functional and experiential aspect of value. *Journal of Vacation Marketing* 17, 5–15.
Jeong, M. and Shin, H.H. (2019) Tourists' experiences with smart tourism technology at smart destinations and their behavior intentions. *Journal of Travel Research* 59, 1464–1477.
Kansakar, P., Munir, A. and Shabani, N. (2019) Technology in the hospitality industry: prospects and challenges. *IEEE Consumer Electronics Magazine* 8, 60–65.
Kheiri, J. and Nasihatkon, B. (2015) Evaluating the effects of social media usage on tourist's behavior based on different phases of travel process. *SIT Journal of Management* 5, 21–40.
Kim, H., Woo, E. and Uysal, M. (2015) Tourism experience and quality of life among elderly tourists. *Tourism Management* 46, 465–476.
Kim, M. and Thapa, B. (2018) Perceived value and flow experience: application in a nature-based tourism context. *Journal of Destination Marketing & Management* 8, 373–384.
Kim, S., Ham, S., Moon, H., Chua, B.L. and Han, H. (2019a) Experience, brand prestige, perceived value (functional, hedonic, social, and financial), and loyalty among GROCERANT customers. *International Journal of Hospitality Management* 77, 169–177.

Kim, Y., Kim, C.K., Lee, D.K., Lee, H.W. and Andrada, R.I.T. (2019b) Quantifying nature-based tourism in protected areas in developing countries by using social big data. *Tourism Management* 72, 249–256.

Li, S.C., Robinson, P. and Oriade, A. (2017) Destination marketing: the use of technology since the millennium. *Journal of Destination Marketing & Management* 6, 95–102.

Loureiro, S.M.C., Stylos, N. and Miranda, F.J. (2020) Exploring how mindfulness may enhance perceived value of travel experience. *The Service Industries Journal* 40, 800–824.

Luis, J.J.G., Cameron, B., Crawley, E. and Net, M.S. (2018) System architecture for tracking passengers inside an airport terminal using RFID. *2018 IEEE Aerospace Conference*, pp. 1–16.

Lykou, G., Anagnostopoulou, A. and Gritzalis, D. (2019) Smart airport cybersecurity: threat mitigation and cyber resilience controls. *Sensors* 19, 19.

Mallick, R., Kilari, K., Shreya, A.S. and Kiranmayee, T.S. (2018) IoT based airport baggage tracing system. *Journal of Network Communications and Emerging Technologies* 8(4).

Marlina, E. and Natalia, D.A. (2017) Land transportation and tourism development. *International Journal of Economic Perspectives* 11, 542–554.

Mayer, C. (2019) Digital passengers: a great divide or emerging opportunity? *Journal of Airport Management* 13, 335–344.

Meng, L., Somenahalli, S. and Berry, S. (2020) Policy implementation of multi-modal (shared) mobility: review of a supply-demand value proposition canvas. *Transport Reviews* 40, 670–684.

Mohseni, S., Jayashree, S., Rezaei, S., Kasim, A. and Okumus, F. (2018) Attracting tourists to travel companies' websites: the structural relationship between website brand, personal value, shopping experience, perceived risk and purchase intention. *Current Issues in Tourism* 21, 616–645.

Neuhofer, B., Buhalis, D. and Ladkin, A. (2012) Conceptualising technology enhanced destination experiences. *Journal of Destination Marketing & Management* 1, 36–46.

Neuhofer, B., Buhalis, D. and Ladkin, A. (2015) Smart technologies for personalized experiences: a case study in the hospitality domain. *Electronic Markets* 25, 243–254.

Novak, T.P. and Hoffman, D.L. (2018) Relationship journeys in the internet of things: a new framework for understanding interactions between consumers and smart objects. *Journal of the Academy of Marketing Science* 47, 216–237.

Oriade, A. and Schofield, P. (2019) An examination of the role of service quality and perceived value in visitor attraction experience. *Journal of Destination Marketing & Management* 11, 1–9.

Pandža Bajs, I. (2015) Tourist perceived value, relationship to satisfaction, and behavioral intentions: the example of the Croatian tourist destination Dubrovnik. *Journal of Travel Research* 54, 122–134.

Patel, D., Narmawala, Z., Tanwar, S. and Singh, P.K. (2019) A systematic review on scheduling public transport using IoT as tool. *Smart Innovations in Communication and Computational Sciences* 4, 39–48.

Pesonen, J. and Horster, E. (2012) Near field communication technology in tourism. *Tourism Management Perspectives* 4, 11–18.

Pestek, A. and Sarvan, M. (2020) Virtual reality and modern tourism. *Journal of Tourism Futures*.[ahead-of-print: doi.org/10.1108/JTF-01-2020-0004].

Pham, Q.T., Tran, X.P., Misra, S., Maskeliūnas, R. and Damaševičius, R. (2018) Relationship between convenience, perceived value, and repurchase intention in online shopping in Vietnam. *Sustainability* 10, 156.

Pillmayer, M., Scherle, N. and Volchek, K. (2021) Destination management in times of crisis: potential of open innovation approach in the context of COVID-19? In: Worndl, W., Koo, C. and Stienmetz, J.L. (eds) *Information and Communication Technologies in Tourism*

2021: Proceedings of the ENTER 2021 eTourism Conference, January 19–22. Springer, Cham, Switzerland, pp. 517–529.

Romero, I. and Tejada, P. (2020) Tourism intermediaries and innovation in the hotel industry. *Current Issues in Tourism* 23, 641–653.

Salman, T. and Jain, R. (2019) A survey of protocols and standards for internet of things. *Advanced Computing and Communications* 1(1).

Sánchez-Fernández, R. and Iniesta-Bonillo, M.Á. (2007) The concept of perceived value: a systematic review of the research. *Marketing Theory* 7, 427–451.

Shaheen, S. and Cohen, A. (2018) Is it time for a public transit renaissance? Navigating travel behavior, technology, and business model shifts in a brave new world. *Journal of Public Transportation* 21, 8.

Sharma, A., Sharma, S. and Chaudhary, M. (2020) Are small travel agencies ready for digital marketing? Views of travel agency managers. *Tourism Management* 79, 104078.

Sheth, J.N., Newman, B.I. and Gross, B.L. (1991) *Consumption Values and Market Choices: Theory and Applications*. South-Western Publishing Co., Cincinnati, Ohio.

Sodhro, A.H., Obaidat, M.S., Abbasi, Q.H., Pace, P., Pirbhulal, S. *et al.* (2019) Quality of service optimization in an iot-driven intelligent transportation system. *IEEE Wireless Communications* 26, 10–17.

Spachos, P. and Plataniotis, K.N. (2020) BLE beacons for indoor positioning at an interactive IoT-based smart museum. *IEEE Systems Journal* 14, 3483–3493.

Stocker, A. and Shaheen, S. (2018) Shared automated mobility: early exploration and potential impacts. *Road Vehicle Automation* 4, 125–139.

Stylos, N. (2019) Technological evolution and tourist decision-making: a perspective article. *Tourism Review* 75, 273–278.

Stylos, N. and Zwiegelaar, J. (2019) Big data as a game changer: how does it shape business intelligence within a tourism and hospitality industry context? In: *Big Data and Innovation in Tourism, Travel, and Hospitality*. Springer, Singapore, pp. 163–181.

Stylos, N., Zwiegelaar, J. and Buhalis, D. (2021) Big data empowered agility for dynamic, volatile, and time-sensitive service industries: the case of the tourism sector. *International Journal of Contemporary Hospitality Management* 33(3).

Tan, W.K. (2017) The relationship between smartphone usage, tourist experience and trip satisfaction in the context of a nature-based destination. *Telematics and Informatics* 34, 614–627.

Tussyadiah, I.P. and Zach, F.J. (2012) The role of geo-based technology in place experiences. *Annals of Tourism Research* 39, 780–800.

Tussyadiah, I.P., Wang, D., Jung, T.H. and Dieck, M.C.T. (2018) Virtual reality, presence, and attitude change: empirical evidence from tourism. *Tourism Management* 66, 140–154.

Verma, A. and Shukla, V. (2019) Analyzing the Influence of IoT in the tourism industry. *Proceedings of International Conference on Sustainable Computing in Science, Technology and Management*, Amity University Rajasthan, Jaipur, India, pp. 45–68.

Wang, D. and Fesenmaier, D.R. (2013) Transforming the travel experience: the use of smartphones for travel. *Information and Communication Technologies in Tourism* 2013, 58–69.

Wang, X., Li, X.R., Zhen, F. and Zhang, J. (2016) How smart is your tourist attraction? Measuring tourist preferences of smart tourism attractions via a FCEM-AHP and IPA approach. *Tourism Management* 54, 309–320.

Weng, G.S., Zailani, S., Iranmanesh, M. and Hyun, S.S. (2017) Mobile taxi booking application service's continuance usage intention by users. *Transportation Research Part D: Transport and Environment* 57, 207–216.

Wu, H.C. and Cheng, C.C. (2018) Relationships between technology attachment, experiential relationship quality, experiential risk and experiential sharing intentions in a smart hotel. *Journal of Hospitality and Tourism Management* 37, 42–58.

Xiang, Z., Schwartz, Z., Gerdes, J.H. Jr. and Uysal, M. (2015a) What can big data and text analytics tell us about hotel guest experience and satisfaction? *International Journal of Hospitality Management* 44, 120–130.

Xiang, Z., Tussyadiah, I. and Buhalis, D. (2015b) Smart destinations: foundations, analytics, and applications. *Journal of Destination Marketing and Management* 4, 143–144.

Xu, L.D., He, W. and Li, S. (2014) Internet of Things in industries: a survey. *IEEE Transactions on Industrial Informatics* 10, 2233–2243.

Yuksel, A.S., Cankaya, I.A. and Cankaya, S.F. (2020) IoT for hospitality industry: paperless buffet management. In: Information Resources Management Association (ed.) *Securing the Internet of Things: Concepts, Methodologies, Tools, and Applications*. IGI Global, Hershey, Pennsylvania, pp. 1388–1408.

2 Young British Tourists' Tourism-related Information Sources

Ilma Aulia Zaim*

Middlesex University London, UK

2.1 Introduction

Nowadays, tourism is more experience-oriented, influenced not solely by the destination's attractiveness but also by its competitive advantages such as hospitality, accessibility, infrastructure and other supporting factors (Ritchie and Crouch, 2010; Fernando and Long, 2012; Zehrer *et al.*, 2017; Chin *et al.*, 2017). Destination image is a component that plays a significant part in driving tourism destination competitive advantages (Konecnik, 2002; Ritchie and Crouch, 2010; Fernando and Long, 2012), since positive destination images affect tourists' experiences, satisfaction and loyalty (Kiráľová and Pavlíčeka, 2015; Niu *et al.*, 2019).

Destination image formation is based on perceptions or impressions generated from a flood of information (Echtner and Ritchie, 1991; Frías *et al.*, 2012; Llodrà-Riera *et al.*, 2015). The flood of information can have many sources, namely family, friends or travel agents' opinions, promotional or advertising media such as travel brochures and posters, and general media including newspapers, TV programmes, magazines, books and films (Pan *et al.*, 2014, 2017; Llodrà-Riera *et al.*, 2015; Terzidou *et al.*, 2017). Each destination image-formation agent has its own unique attributes that influence tourists' decision-making process prior to visiting a destination. For example, social media have become a primary information source for many tourists as they allow experience sharing and provide a credible information source to plan the travel (Govers *et al.*, 2007; Xiang and Gretzel, 2010; Munar and Jacobsen, 2014). Accordingly, tourists' preferences of tourism information sources are an important aspect to be further understood by Destination Management Organizations (DMOs) and tourism marketers to manage the proper information channels for specific tourist segmentations.

*ilma.zaim@gmail.com

DOI: 10.1079/9781789248753.0002

Tourists from the UK play a very significant part in the global tourism market. The European Union reported that the UK is one of the world's top spenders in international tourism, ranked as the fourth largest market after China, the USA and Germany (World Tourism Organization, 2018). According to the report, despite the British sterling depreciation and Brexit, UK residents' visits to international tourism destinations grew 8% in 2018 to 69 million, with tourist spending up to 59 billion euros. However, since international tourism declined due to the Covid-19 pandemic, the UK's expenditure decreased by 89.3% in 2020 (World Tourism Organization, 2020).

According to UNWTO (2016), young travellers coming from two generation cohorts (Generation Z and millennials) are those that represent the rapidly growing segment of international tourism with high value for the destinations, since they tend to travel for a long time and spend more on average compared to older travellers. Therefore, it is essential to understand young British tourists' travel information source preferences to prepare for the resurgence of international tourism in the post-pandemic era. Through in-depth interviews with 25 Londoners aged 18–34, this chapter discusses how the young British search for tourism information that helps their travel decision-making before travelling to destinations and provides DMOs and tourism marketers with insights into their information source preferences.

2.2 Tourism Information Sources

Gunn (1972) identified two types of destination image based on information sources used: (i) organic image, for example when all the information is unintentionally delivered by non-tourism representatives such as via TV, books or friends' recommendations or by those people living at the tourists' destination; and (ii) induced image, where the image is formed by the promotional effort or publicity of the tourist destination. As an addition to Gunn's theory, Gartner (1994) suggested eight agents that work to form destination image: (i) Overt Induced I (traditional types of advertising such as TV, radio and brochures); (ii) Overt Induced II (information from tour operators or tourism organizations to create attractive destination images for certain areas); (iii) Covert Induced I (well-known spokespersons or celebrities via traditional forms of advertising); (iv) Covert Induced II (tours or media groups for travel writers who contribute travel articles in newspapers); (v) Autonomous (independent movies, documentaries or news reporting); (vi) Unsolicited Organic (unrequested information received from individuals who have visited a destination); (vii) Solicited Organic (information received from friends or relatives); and (viii) Organic (actual personal experiences from the trip). With the massive development of digital information and communication technology nowadays, some information sources might have been shifted to different roles, which influence the formation of destination image. Therefore, investigating tourists' preferences of information sources is a pivotal step to understanding the role of destination image agents in this digital era.

A number of researchers have examined tourists' information search behaviour (e.g. Almeida-Santana *et al.*, 2018; De la Hoz-Correa and Muñoz-Leiva, 2019; Kolar, 2017; Thompson *et al.*, 2017). Almeida-Santana *et al.* (2018), for example, used a computer-assisted web of interviews and surveys of tourists from 17 European countries to investigate the European tourist market segmentations through information sources and motivation. Their research categorized three groups of tourists based on the information sources used. For instance, one group of tourists was labelled 'ensurers with digital exploration' as they extensively use YouTube, Wikipedia and Facebook as well as Tripadvisor as tourism information sources, while the other two were named 'ensurers' and 'low social media users'. However, despite the plethora of studies on tourists' information searches, there is still a lack of understanding of how specific age groups use information sources to access tourism information.

Other studies have also stressed the role of organic information sources such as social media, travel forums, travel blogs, tourist photography and tourists' videos in destination image formation. For example, Chung and Koo (2015), Berhanu and Raj (2020) and Mariani *et al.* (2019) discussed the role of social media and argued that they significantly influence perceived value and travel decision-making. Researchers (e.g., Cortez *et al.*, 2014; Liu-Lastres *et al.*, 2015; Tamajon and Valiente, 2017) who studied the role of e-Word of Mouth found that tourists consider online destination reviews, including comments, rankings and ratings on Tripadvisor, very useful to acquire information about a destination. Some others (e.g. Tussyadiah and Fesenmaier, 2009; Dinhopl and Gretzel, 2016; Lodha and Philip, 2019) explored the role of tourists' videos and concluded that tourist video mediates tourist experience and influences the decision to plan a trip. Given the multi-faceted features of organic agents, it is essential to further investigate how tourists value each attribute of the information source when looking for tourism information.

Despite numerous studies on tourism information sources, the young tourist's information-seeking behaviour and the process of information search are still under-researched in the tourism literature, offering opportunities for further research. Understanding the younger generation's attitude and tourism-related behaviour is essential study for tourism researchers since young tourists' demands, needs and wants will shape the future of the tourism industry (Caber *et al.*, 2020). Moreover, youth tourism offers huge opportunities for international tourism post-Covid 19 (Richards and Morrill, 2020). Given the important role of the British outbound market for destination growth in global tourism, this study will focus on young British tourists.

2.3 Methodology

This research used a qualitative approach to explore British tourists' preferences in tourism information sources when seeking travel inspirations, recommendations or guidance before visiting a destination. Accordingly, this study lies in the constructivism paradigm, as it facilitates the construction of knowledge through human interaction (Saunders *et al.*, 2019).

At least 96% of households in Great Britain access the internet with at least 15.7 million Britons aged 16–34 dominating the internet population, and over 35% of all British adults use 'internet-of-things' devices (Office for National Statistics UK, 2020). Given the fact that the internet and ICT have reformed the distribution of tourism information and the way people plan and travel (Xiang and Gretzel, 2010; McDool *et al.*, 2020), this study designates British internet users aged 18–34 as study participants. Internet users under 18 were not involved in this study as they are still under parental responsibility (Parliament UK Children Law and Practice, 2017).

This study conducted one-on-one semi-structured in-depth interviews with 25 respondents (House of Commons Library, 2020). The interviews were conducted at the British Library, London. The participants were recruited through callforparticipants.com, a UK-based platform that enables researchers to advertise their surveys, interviews or experiments to potential participants. The researcher also provided the interested respondents with a Google Form link to select the appointment date, collect the participants' demographic characteristics and ensure they met all the research criteria.

Questions such as 'Do you conduct any research before travelling?', 'What kind of sources do you use to get information about a destination?' and 'Which one do think offers more reliable information?' were asked to gain insights on how British tourists search for tourism-related information and were followed by further questions to get an idea of their travel profiles. All the interviews were transcribed by the researcher herself, verbatim, by including all verbal and non-verbal expressions, such as pauses, laughter, sighs, etc., to capture the original nature of the interview (Braun and Clarke, 2006). The analysis was done through thematic analysis using N-VIVO version 12. The next section discusses the findings of the interviews including the demographic characteristics and the travel profile of the study participants.

2.4 Multiple Sources of Information

The total sample of the study was 25 British tourists aged 18–34. Regarding gender, 46% were male and 52% were female. It appears that the majority of the British individuals who took part in this study were students (56%), followed by employees (24%), self-employed (12%) and unemployed (8%).

In terms of travel profile, at least 72% of the participants travel up to five times a year for work or holiday purposes. All the participants have travelled to some domestic destinations around the UK in the last three years and over two thirds have travelled overseas in the same period. European countries such as France, Spain, Germany and Italy were the most visited destinations for British travellers, 'for spending weekends'. Meanwhile, about a third of participants said they travelled during a gap year when they left school. About a quarter of the interviewees also liked solo travelling and stressed 'freedom' as their reason for travelling alone.

Prior to the actual visit to the destination, British tourists typically seek tourism-related information using five primary sources: (i) suggested links via Google; (ii) YouTube; (iii) social networking sites (Instagram, Facebook and Snapchat); (iv) personal sources of information; and (v) travel guidebooks. These information sources play a significant part in British tourists' decision-making processes and influence destination image formation. All the sources of information are connected and constitute a sequential information-seeking pattern.

The interviews reveal that British tourists actively search for tourism-related information using multiple sources. The process of information search typically involves more than two sources, which are followed by other sources. For example:

> 'I use websites, maybe go through blog posts, or maybe even read, like, holiday.com and things like that. And then I look through Instagram, you know, through the hashtags of that specific country and just see what comes up.' (BRIT-21, female, 20)

The following section discusses each source of information used by British tourists.

2.4.1 Suggested links provided by Google

It appears that Google is the number-one source of information for young tourists from the UK. A significant number of participants use Google in their initial information search before checking for further information on other internet platforms:

> 'I often Google stuff, or maybe Google stuff to find, like, names of places, and then look at Instagram pictures and things like that.' (BRIT-08, male, 22)

British tourists argued that Google provides 'suggested links' based on the keywords they typed. Google also suggests extensive information across internet platforms; namely images, maps, websites, travel forums, blogs, travel agents and videos. These young British tourists also use Wikipedia to get detailed information about the destination. For example, a participant said:

> 'I usually read, like, the Wikipedia page briefly. I'll be, like, I might find interesting population and GDP things, like, political and economic stuff, like socioeconomic aspects, environmental things – stuff like that.' (BRIT-13, male, 21)

From Google, young British tourists go to Tripadvisor to get further information. The interviews indicate that 'visitors' reviews' are the reason why a dozen participants use Tripadvisor to find information about various tourism attributes, such as hotels, restaurants, activities or attractions. They value Tripadvisor for its authenticity, as it is perceived as a trustworthy tourism information source that they genuinely rely upon. As one participant said:

> 'There's a varied amount of information you can find, and I just feel like it's so much better when you hear from normal people, not influencers or bloggers that maybe have paid advertisements. It's just for normal people and you just have their opinions and I feel like that's more genuine for me.' (BRIT-28, female, 22)

Social media have generated a new type of celebrity, namely micro-celebrities, who can create engagement with their followers and who have become increasingly used by marketers to reach their target markets on social media, defined as influencer marketing (Gretzel, 2018). While influencers demonstrate pseudo-events (Boorstin, 1961; Abidin, 2016), Tripadvisor offers greater authenticity from actual tourists' reviews, which generate a more trustworthy image of the destination (Sumarsono *et al.*, 2019). About one quarter of Tripadvisor users use this platform as the first phase of their research, then go to other platforms such as YouTube or Instagram that provide visual features to get a better idea of the destination. For example:

> 'I always use Tripadvisor to look at specific places…and then if from Tripadvisor I know there's a restaurant I want to go to, I go on Instagram or find it on Facebook to look at the pictures of the food.' (BRIT-11, female, 28)

Via Google, a small group of British tourists go to blogs to get detailed information about tourist destinations because they offer travel photos, trip stories and travel recommendations based on 'personal experience'. For example, 'I would look at blog posts about Bali stuff, like the waterfall jumping' (BRIT-09, female, 21). A larger number of British interviewees, by contrast, revealed that they prefer blogs less:

> 'I don't really look for blogs; I think it's, like, articles that someone has posted. But if the blog is there, I click on that but not specifically look for blogs.' (BRIT-06, female, 21)

This is somewhat in line with Thompson *et al.* (2017) who argued that British tourists rarely use blogs as an information source compared to other sources such as Tripadvisor.

The interviews also show evidence that tourism service providers help British tourists to plan their travel budgets and trip itinerary. About one tenth of British participants, especially those between 18 and 30, also reveal tourism service providers that provide flight and accommodation bookings as part of their information-seeking process. Furthermore, these participants highlighted *affordability, with reference to* flight and accommodation prices, as the main factor. As illustrated: *'The only information I gather is: Is the food affordable? How much is the accommodation? Do I have enough?'* (BRIT-15, male, 28). A British participant described this process as a *'logistical approach'* to trip planning:

> 'Yeah [I go to tourism service providers] most of the time, because of being a student, I usually buy my flights, like, in advance – it's so much cheaper

...Yeah, so it's, I'd definitely say, it's like a logistical, a logical, organized approach to planning...Yes, flights are usually just on Skyscanner because it's super flexible.' (BRIT-13, male, 21)

Travellers, especially students, compare prices among hotels and airlines when planning a trip. Price is the main factor for student travellers, which influences their decision-making (Bicen and Sadikoglu, 2016); for example, they mostly prefer low-cost airlines over traditional carriers (Gheorghe *et al.*, 2017). Mobile apps such as Skyscanner and Booking.com have a significant impact on students' travel-planning decisions (Bicen and Sadikoglu, 2016).

2.4.2 YouTube

YouTube is ranked as the second most popular travel-related information source used by the British. British participants also stated that they usually watch videos to get a more realistic view of the destination:

'The most reliable platform for me I would say is YouTube. Yeah, YouTube videos...Because particular person travelled there, and it's good to see their views and how is it actually like there, so you can see the real situation, and you feel like there's another person like you, especially if you, like, relate to them and stuff...For example if they stayed in a hotel or like, certain places they visited, and you kind of see the place and you kind of drop that down and keep note.' (BRIT-21, female, 20)

All participants use YouTube for watching videos, but only two of them have a YouTube account, although they watch videos on the site almost every day. A total of 30 British participants use YouTube, confirming a Digital in the UK Report (2019) that YouTube is the most active social media platform in the UK. It is reported that at least 80% of UK internet users use this platform. The use of mobile devices has a positive impact on British social media usage (Blank and Lutz, 2017).

The vast majority of British participants use YouTube for watching travel vlogs. These interviewees expressed their positive views regarding YouTube videos as a useful reference for holiday destinations. For instance:

'Oh yes, I love watching travel videos on YouTube...I mean like videos produced by the tourists who've been to the destination...It's very good, because, for example, it's another sort of source of information of particular destination you want to go and that is an opportunity for you to have an idea of the place...like...if I receive, like, a message from someone, hey check out this place...I would look at that...you'll see the environment... like, how the buildings are like there...things like that.' (BRIT-18, male, 34)

It is also revealed that British tourists go to YouTube as they seek 'visual confirmation' after receiving information from other platforms. This platform is used as the first step of the information-seeking process:

'YouTube. Super-useful. I saw loads of, like, vlogs by people who've been to Bali...Yeah, YouTube, just because you can see a real video. You can see what

> it's like in real time rather than just the blog being written about it, then you have to imagine it.' (BRIT-09, female, 21)

YouTube users use the platform for seeking tourism information (Reino and Hay, 2011). YouTube offers numerous travel videos, which give a richer experience and are more effective at engaging a person's feelings and perceptions compared to a text or photograph, because the spatial features of videos enable the viewer to feel closer to the place and the direct experience, as well as to gain a more realistic feeling of actually being there (Marlow and Dabbish, 2014). Shared travel videos on YouTube are mediators of tourist experiences as they evoke mental pleasures and help to construct fantasies and daydreams of visiting particular destinations (Tussyadiah and Fesenmaier, 2009).

2.5 Social Networking Sites (SNS)

In their discussion on social media categories, Kaplan and Haenlein (2010) distinguish between social networking sites (e.g. Facebook) and content communities (e.g. YouTube). Although both types are categorized as social media, each has its own features and characteristics. Social networking sites (SNS) are platforms that enable users to connect with other users by creating personal profiles, sending messages and sharing personal information including photos, videos or blogs. Content communities platforms enable their users to share specific media content, for example videos, on YouTube (Kaplan and Haenlein, 2010).

Most participants have more than one social media account. About one third of British participants use more than three social media. With the vast development of information technology, nowadays people can access a number of different media at the same time (Madianou and Miller, 2012; May and Elder, 2018). As documented from the interviews, the three top social networking sites that all the participants use are Facebook, Instagram and Snapchat.

Unlike other types of social networking sites, Instagram is mainly used as the first stage of the information search process concerning tourism destinations. The young British valued Instagram for its 'story' feature, which enables its users to see the real-time image of the destination from the people who are experiencing it. Apart from that, participants responded that compared to other social media platforms, Instagram offers *'geotagging'* and *'explore'* features that make the information search process easier. For instance:

> 'So I search by, like, a location, geotagging, to see the poses that other people have taken, or to get a better visual understanding of the place, to kind of see where I could possibly go, and maybe, like, borrow some interesting angles and things like that.' (BRIT-13, male, 21)

One tenth revealed Facebook as an 'information source', as it features numerous travel groups that enable travellers from all around the world to share their travel experiences or to ask travel-related questions. Facebook also allows people to share their common interests and contact pages for communicating with companies or service operators to get further information on what is being offered. As illustrated:

> 'On Facebook, there is a group called Culture Trip, and they post some cool videos about destinations…and because, as backpackers, we have our own kind of community, so when I was in Australia, the Australian backpackers would post stuff on interesting areas, buy and sell items like camping cars, so we're enlightened. We get tips about tax or things like that; we swap information and message that person.' (BRIT-01, male, 30)

One third of participants use Snapchat. British participants use this platform as it offers intimacies through the temporality and the short duration of photo-sharing (Kofoed and Larsen, 2016; Handyside and Ringrose, 2017; Jeffrey *et al.*, 2019). Nonetheless, the temporariness of photo-sharing on Snapchat has driven its users to post pictures with less concern for the quality, unlike posts on Instagram or Facebook, which are typically planned and polished (Kofoed and Larsen, 2016; Jeffrey *et al.*, 2019). Hence, Snapchat allows its users to see the actual image of a place from friends who are at the destination:

> 'Before I jump onto Google, I'm going to speak to that person…So, for example, I saw my friend Snapchatted and so she went to stay in a house in Wales…I saw the video and I was like, oh my God, that looks like such a nice place, I want to go. So I said where is this? So I have her in conversation on Snapchat, then when I'm ready to go, I want to go to Wales.' (BRIT-02, female, 28)

It has been affirmed that social media plays a part as a source of information for tourists and influences travel decision-making (e.g. Chung and Koo, 2015; Qian *et al.*, 2015). However, although social media users value social media for its role as an information source, the majority of participants understand that social networking sites also extensively expose visual materials that do not depict the physical reality of the place because they have been edited to make them look more appealing (Kim and Stepchenkova, 2015). As stated: 'My fiancé and I booked a hotel because the pictures looked very nice, but when we got there it was the worst hotel ever' (BRIT-11, female, 28).

Tourism scholars have confirmed that there are discrepancies among induced, autonomous and organic images (e.g. Stepchenkova and Zhan, 2013; Terzidou *et al.*, 2017; Mak, 2017). Accordingly, non-induced information delivered by actual visitors is a prominent tool that enables potential tourists to obtain a more accurate image of the place (Gartner, 1994; Liu-Lastres *et al.*, 2015). With the vast development of smartphone- and internet-based social media, tourists not only share their experience in the form of digital photography but also in online videos (Tussyadiah and Fesenmaier, 2009). This study shows evidence that young British SNS users go to YouTube to see tourists' videos portraying the real image of the destination.

2.6 Personal Sources of Information

It is revealed that less than half a dozen use personal sources of information as they value 'first-hand experience' and 'specific interest' from those who have visited the destination, especially if they are friends, family or relatives. As one interviewee said:

> 'Well, I'm always going to trust first-hand experience. So, my friends, my family who've gone to the place...My friends and family, it's almost like a filter of what I would like.' (BRIT-01, male, 30)

The findings show that the credibility of personal sources of information influences trustworthiness. This aligns with studies that suggest friends' opinions influence decision-making, such as hotel bookings (Gellerstedt and Arvemo, 2019), holiday planning (Thompson *et al.*, 2017) or hospital selection for medical tourism (Yeoh *et al.*, 2013). The prominent role of personal sources of information in the travel decision-making process has been extensively discussed in tourism literature (e.g. Murphy *et al.*, 2007; Gellerstedt and Arvemo, 2019). However, in this digital information technology era, personal sources of information from friends, family or relatives are not usually considered the most important sources of information. The majority of participants rely more on e-WOM such as blogs or Tripadvisor, or social networking sites and YouTube as these online platforms provide information that is easier and faster to access (Huete-Alcocer, 2017). Yet despite the vast development of digital technology, a quarter of the participants use travel guidebooks to find the information they seek.

2.7 Travel Guidebooks

This study uncovers 'travel guidebooks' as information sources that are particularly used by young British tourists. One fifth of participants disclosed their reliance on travel guidebooks as a source to get travel recommendations. Some of these participants mentioned 'Lonely Planet' and 'Eyewitness Travel'. An interviewee claimed:

> '...when I was to go to China and India. I went there for a month, so I had a book and then I had all kinds of information in that.' (BRIT-06, female, 21).

A travel guidebook such as Lonely Planet is 'the bible' (Hanlan and Kelly, 2005, p. 169) for tourists from the UK. However, the existence of the internet has increasingly replaced travel guidebooks as their information gets rapidly outdated; travellers need fast and updated information while on the move (Young and Hanley, 2010). However, this research finds that despite the dominance of online sources, travel guidebooks are still preferred by British participants aged 18–34. However, it is argued that although travel guidebooks complement the information generated by other sources like WOM

Table 2.1. Summary of findings.

Information sources	Information searched for by young British tourists
Suggested links on Google	Wikipedia: detailed information of destination profiles Tripadvisor: tourists' reviews of hotels, restaurants, etc. Blogs: stories of personal experiences of tourists who have visited the destination Tourism service providers: ticket price, accommodation booking, trip itineraries
YouTube	Visual confirmation of destination images
Social networking sites	Instagram: real-time images through stories, geotag, explores Facebook: travel groups' experience-sharing Snapchat: intimacies through photo-sharing
Personal sources of information	First-hand experience, specific interests from friends, family or relatives who have been to the destination
Travel guidebooks	Travel recommendations

(Hanlan and Kelly, 2005), besides travel guidebooks, participants also use other sources, such as Google, to seek additional information about tourism destinations.

2.8 Conclusion

This chapter discussed how the young British search for tourism information that helps their travel decision-making before the actual visit to the destination and provided DMOs' and tourism marketers' insights on their information source preferences. Through thematic analysis of in-depth interviews of 25 British tourists aged 18–34, resident in London, it was revealed that British tourists typically seek tourism-related information using five primary sources: (i) suggested links via Google; (ii) YouTube; (iii) social networking sites (Instagram, Facebook, Snapchat); (iv) personal sources of information; and (v) travel guidebooks. A summary of the findings is shown in Table 2.1.

Via Google, young tourists from the UK search for initial information before checking for further information on other internet platforms. British tourists use specific Google features such as Google Images to get information on the destination. Through the keywords of the information they seek, they also use suggested links provided by the search engine, for example Wikipedia, Tripadvisor, blogs and also tourism service providers. Given the high dependence of young British tourists on Google and some specific links, this study suggests DMOs and tourism marketers drive clicks through the search engine results page so that their websites could stand out at the top of the Google page. Studies that discussed the search engine implications for tourism marketing (e.g. Xiang and Pan, 2011; Pan, 2015) argued that being at the top of the page gives an advantage and makes the business more visible to the tourist. Therefore, business or DMOs should build relevant keywords,

URLs, snippets and attractive web content to attract more clicks so that they can appear at the top (Pan, 2015).

After Google, YouTube is ranked as the second most important travel-related information source used by the British. It is revealed that young British tourists usually watch videos to get a more realistic view of the destination through travel vlogs. These British tourists also go to YouTube to seek 'visual confirmation' after receiving information from other platforms as they highly value image authenticity. YouTube offers numerous travel videos, which give a richer experience and are more effective at engaging a person's feelings and perceptions compared to a text or a photograph, because the spatial features of videos enable the viewer to feel closer to the place and the direct experience, as well as to gain a more realistic feeling of actually being there (Marlow and Dabbish, 2014). Accordingly, DMOs and tourism marketers should create commercial YouTube videos that allow tourists to see the real image of the destination, to generate richer tourist experiences and interactivity through information-seeking on YouTube (Tiago *et al.*, 2019).

The interviews also indicated that most young British tourists have more than one social networking site account. As seen from the interviews, most British participants use Facebook, followed by Instagram and Snapchat. Although British tourists value social networking sites for their role as an information source, the majority of participants understand that these platforms also extensively expose visual materials that do not depict the physical reality of the place because they have been edited to make them look more appealing (Kim and Stepchenkova, 2015). As such, in order to improve young British tourists' trust in the travel information on social media, this study suggests DMOs and tourism marketers should create content with more authentic photos or videos by involving their young followers in sharing their travel experiences, to be reposted on the destination's social media channels through relevant hashtags.

Next, findings show that British tourists use word-of-mouth as a solicited organic source, as they value 'first-hand experience' and 'specific interest' from those who have visited the destination, especially if they are friends, family or relatives. British tourists appreciate the role of WOM as being trustworthy, as it comes from people who have actual experience of visiting the place in question. Thus, DMOs should improve the destination image attributes to drive positive tourist experiences in order to generate positive word-of-mouth (Gellerstedt and Arvemo, 2019). From a marketing point of view, DMOs also need to assure congruity between projected and perceived images to avoid misleading, which might result in disappointment and negative word-of-mouth (Michalkó *et al.*, 2015; Van Es and Reijnders, 2016).

Lastly, despite the enormous number of online information sources, it appeared that British tourists aged 18–34 still rely on travel guidebooks such as Lonely Planet and Eyewitness Travel as the source of travel recommendations, especially when they have already decided to visit the destination. Since travel guidebooks are information sources that support images generated from other sources (Hanlan and Kelly, 2005), DMOs can take advantage

House of Commons Library (2020) Children: parental responsibility – how it's gained and lost, and restrictions (England and Wales). Available at: https://commonslibrary.parliament.uk/research-briefings/cbp-8760/ (accessed 11 June 2021).

Huete-Alcocer, N. (2017) A literature review of word of mouth and electronic word of mouth: implications for consumer behavior. *Frontiers in Psychology* 8, 1256. DOI: 10.3389/fpsyg.2017.01256.

Jeffrey, H., Ashraf, H. and Paris, C. (2019) Hanging out on Snapchat: disrupting passive covert netnography in tourism research. *Tourism Geographies* 23, 144–162.

Kaplan, A.M. and Haenlein, M. (2010) Users of the world, unite! The challenges and opportunities of social media. *Business Horizons* 53, 59–68.

Kim, H. and Stepchenkova, S. (2015) Effect of tourist photographs on attitudes towards destination: manifest and latent content. *Tourism Management* 49, 29–41.

Kiráľová, A. and Pavlíčeka, A. (2015) Development of social media strategies in tourism destinations. *Procedia – Social and Behavioral Sciences* 175, 358–366.

Kofoed, J. and Larsen, M. (2016) A snap of intimacy: photo-sharing practices among young people on social media. *First Monday* 21(11).

Kolar, T. (2017) Exploring combined tourist offerings: the case of sightrunning. *Current Issues in Tourism* 3500, 1–9.

Konecnik, M. (2002) The image as a possible source of competitive advantage of the destination – the case of Slovenia. *Tourism Review* 57, 6–12.

Liu-Lastres, B., Pennington-Gray, L., Donohoe, H.M. and Omodior, O. (2015) New York City bed bug crisis as framed bt tourists on Tripadvisor. *Tourism Analysis* 20, 243–250.

Llodrà-Riera, I., Martínez-Ruiz, M., Jiménez-Zarco, A. and Izquierdo-Yusta, A. (2015) A multidimensional analysis of the information sources construct and its relevance for destination image formation. *Tourism Management* 48, 319–328.

Lodha, R. and Philip, L. (2019) Impact of travel blogs and vlogs on the decision making among the students of Bangalore. *International Journal of Scientific Research and Review* 7, 1060–1076.

Madianou, M. and Miller, D. (2012) Polymedia: towards a new theory of digital media in interpersonal communication. *International Journal of Cultural Studies* 16, 169–187.

Mak, A. (2017) Online destination image: comparing national tourism organisations' and tourists' perspectives. *Tourism Management* 60, 280–297.

Mariani, M., Styven, M. and Ayeh, J. (2019) Using Facebook for travel decision-making: an international study of antecedents. *International Journal of Contemporary Hospitality Management* 31, 1021–1044.

Marlow, J. and Dabbish, L. (2014) Computers in human behavior: when is a picture not worth a thousand words? The psychological effects of mediated exposure to a remote location. *Computers in Human Behavior* 30, 824–831.

May, K. and Elder, A. (2018) Efficient, helpful, or distracting? A literature review of media multitasking in relation to academic performance. *International Journal of Educational Technology in Higher Education* 15, 13.

McDool, E., Powell, P., Roberts, J. and Taylor, K. (2020) The internet and children's psychological wellbeing. *Journal of Health Economics* 69, 102274. DOI: 10.1016/j.jhealeco.2019.102274.

Michalkó, G., Irimiás, A. and Timothy, D. (2015) Disappointment in tourism: perspectives on tourism destination management. *Tourism Management Perspectives* 16, 85–91.

Munar, A. and Jacobsen, J. (2014) Motivations for sharing tourism experiences through social media. *Tourism Management* 43, 46–54.

Murphy, L., Mascardo, G. and Benckendorff, P. (2007) Exploring word-of-mouth influences on travel decisions: friends and relatives vs. other travellers. *International Journal of Consumer Studies* 31, 517–527.

Niu, Y., Park, H. and Kirilenko, A. (2019) Theme park visitor experience and satisfaction: a case of TripAdvisor reviews of three theme parks in Orlando. *Travel and Tourism Research Association: Advancing Tourism Research Globally* 11.

Office for National Statistics UK (2020) Internet access: households and individuals, Great Britain 2020. Available at: https://www.ons.gov.uk/peoplepopulationandcommunity/householdcharacteristics/homeinternetandsocialmediausage/bulletins/internetaccesshouseholdsandindividuals/2020#:~:text=In%20January%20to%20February%202020,2006%20when%20comparable%20records%20began (accessed 11June 2021).

Pan, B. (2015) The power of search engine ranking for tourist destinations. *Tourism Management* 47, 79–87.

Pan, S., Lee, J. and Tsai, H. (2014) Travel photos: motivations, image dimensions, and affective qualities of places. *Tourism Management* 40, 59–69.

Pan, S., Santos, C. and Kim, S. (2017) Promoting tourism, projecting power: the role of television commercials. *Journal of Travel and Tourism Marketing* 34, 192–208.

Parliament UK Children Law and Practice (2017) House of Commons Library. Available at: https://commonslibrary.parliament.uk/research-briefings/sn02827/ (accessed 5 February 2017).

Qian, Y., Hu, J. and Zhang, M. (2015) Study on the online travel information search: a perspective of social media. *Advances in Applied Sociology* 5, 219–233.

Reino, D. and Hay, B. (2011) The use of YouTube as a tourism marketing tool. *Proceedings of the 42nd Annual Travel & Tourism Research Association Conference, London, Ontario, Canada.* Travel & Tourism Research Association.

Richards, G. and Morrill, W. (2020) The impact and future implications of COVID-19 in the youth travel sector. *ATLAS Tourism and Leisure Review* 2, 57–64.

Ritchie, J. and Crouch, G. (2010) A model of destination competitiveness/sustainability: Brazilian perspectives. *Revista de Administração Pública* 44, 1049–1066.

Saunders, M., Lewis, P. and Thornhill, A. (2019) *Research Methods for Business Students*, 8th edn. Pearson.

Stepchenkova, S. and Zhan, F. (2013) Visual destination images of Peru: comparative content analysis of DMO and user-generated photography. *Tourism Management* 36, 590–601.

Sumarsono, D., Sudardi, B. and Abdullah, W. (2019) The influence of TripAdvisor application usage towards hotel occupancy rate in Solo. *Journal of Physics: Conference Series* 1175, 012248.

Tamajon, L. and Valiente, G. (2017) Barcelona seen through the eyes of TripAdvisor: actors, typologies and components of destination image in social media platforms. *Current Issues in Tourism* 20, 33–37.

Terzidou, M., Stylidis, D. and Terzidis, K. (2017) The role of visual media in religious tourists' destination image, choice, and on-site experience: the case of Tinos, Greece. *Journal of Travel and Tourism Marketing* 35, 306–319.

Thompson, M., Cassidy, L., Prideaux, B., Pabel, A. and Anderson, A. (2017) Friends and relatives as a destination information source. *Advances in Hospitality and Leisure* 13, 111–126.

Tiago, F., Moreira, F. and Borges-Tiago, T. (2019) YouTube videos: a destination marketing outlook. In: Kavoura, A., Kefallonitis, E. and Giovanis, A. (eds) *Strategic Innovative Marketing and Tourism*. Springer, pp. 877–884.

Tussyadiah, I. and Fesenmaier, D. (2009) Mediating tourist experiences: access to places via shared videos. *Annals of Tourism Research* 36, 24–40.

UNWTO (2016) *The Global Report on the Power of Youth Travel*. UNWTO, Madrid.

Van Es, N. and Reijnders, S. (2016) Chasing sleuths and unravelling the metropolis: analyzing the tourist experience of Sherlock Holmes' London, Philip Marlowe's Los Angeles and Lisbeth Salander's Stockholm. *Annals of Tourism Research* 57, 113–125.

World Tourism Organization (2018) *European Union Tourism Trends*. UNWTO, Madrid.

World Tourism Organization (2020) *World Tourism Barometer*. UNWTO, Madrid.
Xiang, Z. and Gretzel, U. (2010) Role of social media in online travel information search. *Tourism Management* 31, 179–188.
Xiang, Z. and Pan, B. (2011) Travel queries on cities in the United States: implications for search engine marketing for tourist destinations. *Tourism Management* 32, 88–97.
Yeoh, E., Othman, K. and Ahmad, H. (2013) Understanding medical tourists: word-of-mouth and viral marketing as potent marketing tools. *Tourism Management* 34, 196–201.
Young, T. and Hanley, J. (2010) Virtual mobilities: backpackers, new media and online travel communities. The Australian Sociological Association 2010 Conference (TASA 2010). *TASA 2010 Conference Proceedings: Social Causes, Private Lives*, Sydney, 6–9 December.
Zehrer, A., Smeral, E. and Hallmann, K. (2017) Destination competitiveness: a comparison of subjective and objective indicators for winter sports areas. *Journal of Travel Research* 56, 55–66.

3 DMOs and Tourism Marketing Channels in Centre of Portugal

Bernardo Borges and Rui Augusto Costa*

University of Aveiro, Portugal

3.1 Introduction

Tourism is an area where information is a dominant part of all actions and activities. Information and data exchange among all agents involved is essential. Technologies connect all parts in an organic and natural way (Buhalis, 2003).

Digital marketing is, without doubt, one of the most effective ways to promote a tourism destination, creating conditions to communicate efficiently to a huge audience, bringing several benefits to organizations oriented to tourist activity and enabling a greater level of identification of customer needs. Destination management organizations (DMOs) have almost unlimited opportunities through digital technologies to promote destinations and create a clear strategy that can obtain positive results.

The main strategy should involve attracting the attention of tourists, with an optimized use of the various digital channels to obtain feedback, capture attention, shape the image of the destination and disseminate correct and important information. Moreover, a DMO must be integrated in a natural way in the digital world, and constitute a safe information source, enabling the tourist to feel able to obtain everything he/she needs using the various channels made available by the organization.

Several studies over the years have focused on the analysis of digital DMO platforms (Qi *et al.*, 2008; Tanrisevdi and Duran, 2011; Da Silva Pina, 2015; Yang and Wang, 2015; Musina and Gao, 2016; Da Silva Pina *et al.*, 2017; Vyas, 2019), all of them integrating valuable knowledge for the study of strategies promoted by organizations to communicate with tourists. The main objective of this research is to understand whether the official digital marketing channels of Centre of Portugal are being used by tourists in their search for information. This will allow us to understand whether tourists use the

*Corresponding author: rui.costa@ua.pt

DOI: 10.1079/9781789248753.0003

official digital channels of Centre of Portugal in the search for information and whether these channels are indeed important tools for the promotion and communication of destinations in general.

3.2 Destinations and Technologies

3.2.1 Destination management organizations (DMOs)

Several concepts emerged with the integration of technology with tourism, such as the concept of e-tourism (Buhalis, 2003; Werthner and Ricci, 2004), which emerged with the integration of web-based technologies, the integration of social networks, and the recognition of the mobility that tourists have today. This mobility allows tourists to absorb information at almost any time and anywhere (Gretzel *et al.*, 2015).

DMOs are, in general, complex and diverse organizations (Li *et al.*, 2017). According to Pike and Page (2014), a DMO is an organization responsible for the marketing of a destination with a defined geographical area. The same authors state that the focus of the activities of these organizations is on the development of collaborative communication and marketing strategies that combine internal resources (destination) with opportunities in the macro-environment (market). Several authors refute this opinion, considering that, although there is an associated marketing and promotion factor, there are clearly structures that have expanded their perspectives, with competition among destinations associating a management element with organizations (Ritchie and Crouch, 2003; Presenza *et al.*, 2005; Updhya, 2014; Sheehan *et al.*, 2016).

In the new context of destinations, the role of a DMO is no longer just to contribute to the development of new tourism initiatives but to manage human and internal resources in different stages to establish a relationship between the existing infrastructure at the destination and the external market (Minguzzi, 2006). According to Madarasz and Zsofia (2011), DMOs have an increasing importance because competition is no longer carried out among service providers but among destinations themselves.

The truth is that DMOs are in a unique position to be the link between the limits of the destination's internal environment and the external competitive environment, and for that it is necessary to have efficient management capacities. A successful DMO in the future will have to be an agent of the destination, capable of identifying and learning through internal and external stakeholders. It will have to acquire data and filter and analyse it to prioritize information from various sources, and create the understanding that their management capabilities at the destination are efficient (Sheehan *et al.*, 2016).

Clearly, DMOs have a fundamental role in what is the promotion and management of a destination. Managing a destination is complex and diverse. Due to the intrinsic characteristics of the destination, it is necessary to carefully identify who are the stakeholders, to preserve the tourism resources and promote efficient communication. This will create an economically capable destination with a strong image, attracting diverse types of tourists with the capacity to purchase, thereby improving the quality of life of those who live at the destination.

The technological revolution in the sector and the increasing importance of the internet have completely changed the market conditions for DMOs (Buhalis and Law, 2008; Tussyadiah and Fesenmaier, 2009), making them change radically their strategies (Buhalis and Zoge, 2007) and bringing new difficulties to them, for the promotion of destinations (Castro *et al.*, 2017). Technology has enabled the reorganization of the entire tourism system, changing the process of development, management and marketing of tourism products and the destinations themselves (Buhalis *et al.*, 2011).

The digitalization of the travel industry has meant that tourists now have a huge pallet of information, available anywhere and anytime. The more traditional role of the DMO as a means of distributing information to consumers and a structure based on marketing and promotion has changed. In this regard, the management of the destination's image and the way it should be viewed have changed, with the internet playing a vital role. The role of an intermediary, naturally a DMO, does not currently exist, with clients themselves looking for more efficient alternatives in the digital world.

The introduction of new technologies such as social networks and interactive websites, or even virtual reality, facilitates the delivery of information to the tourist, without them needing to travel to the destination and allows them to form an image of the destination without any type of intervention by the DMO in the process (McFee *et al.*, 2019). The existence of online platforms such as Tripadvisor and Facebook has taken the consumer's focus off the promotion or marketing actions spearheaded or co-promoted by the DMO. As such, social networks and digital technologies have become places where DMOs 'fight' daily to get the attention of travellers (Ong and Ito, 2019). In this regard, DMOs have alternative marketing strategies based on the combination of different activities that are inserted in and connected with their websites, these digital tools becoming the main means of disseminating information to tourists and making the tourism product itself more tangible (Morosan, 2008).

Furthermore, consumers are today the co-creators of their experience (Prahalad and Ramaswamy, 2004) and technological trends lead them to be also mediated by technology (Tussyadiah and Fesenmaier, 2009). This makes the action of the DMO more complicated as it needs to adapt its dynamics to emerging trends, with the need to direct the destination to a smarter reality and to become adapted to the technological dynamics the consumer wants. Therefore, DMOs must be able to become relevant in online channels, to have an efficient communication strategy with tourists based on an appropriate marketing strategy. It is necessary to find the best digital marketing tools to communicate optimally with the target audience in order to obtain a constant flow of tourists (Castro *et al.*, 2017).

3.2.2 Digital marketing channels

According to Pomirleanu *et al.* (2013), the importance of digital marketing has grown exponentially and academic interest has increased. For Wymbs (2011), due to the appearance of the internet, various definitions related to marketing and technology have been created, to establish the connection between these areas that are inherently connected.

The concept of digital marketing has undergone notable transformations, due to the constant change in technology and the multiple forms of interaction with the consumer (Castro *et al.*, 2017). According to Chaffey *et al.* (2006), e-marketing includes the use of the internet and other technologies to achieve marketing objectives. Kotler *et al.* (2009) elaborate on this definition, including the term 'interaction', defining digital marketing as being a form of communication and interaction between organizations and their consumers through digital channels and digital technologies. The term interaction is of great importance, with the internet being an essential means of communication today and having a fundamental role in the relationship between organizations and their audiences (Pitana and Pitanatri, 2016). The term is defined by Smith (2011) as the practice of promoting products and services through digital distribution channels.

From these authors it is also possible to realize that digital marketing is much more than just a functional website with an attractive design. Digital marketing is the interconnection between the website, search engines, partner websites, blogs, B2B partners and consumers, with connection being the key. Digital marketing is a process wrapped in technology that aims to collaborate with consumers and partners to create, communicate and deliver value to all stakeholders (Kannan and Li, 2017). In this regard, digital marketing has become an integral part of every industry, regardless of size, characteristics or importance (Kaur, 2017).

According to Oliveira and Panyik (2014), organizations responsible for managing a destination currently have unlimited opportunities to connect with consumers/tourists through different channels and create touchpoints. According to Schmidt *et al.* (2008), tourist organizations increasingly take advantage of the strength of the internet as a vehicle of extreme relevance in the promotion and communication of tourism products and services.

The internet and mobile technologies make it possible to reinforce the position of destinations in the digital world, also reinforcing the online presence and offline connectivity that turns out to be necessary to be competitive in this global market (Adeola and Evans, 2019). Digital marketing and the use of technology allow, above all, quick and dynamic access to the consumer through multiple channels (Leung *et al.*, 2013; Sotiriadis, 2017; Law *et al.*, 2018).

According to Mariani *et al.* (2014), the use of digital channels represents an economic means of promoting a destination. Straker and Wrigley (2016) define them as platforms based on the internet that allow connection of a wide variety of users, and where websites, social networks and blogs are clearly inserted. However, all of them work only if supported by a strategy oriented to the relevant content for the tourist and where the smartphone has extreme importance. The digital channels assigned to a destination serve as safe information sources (Bitzinger, 2018) and must work together to transmit an image of continuity and enable the dissemination of experiences of the destination. Digital tools such as websites, e-mail, blogs and other digital platforms allow dynamic interaction with tourists anytime and anywhere and must be co-ordinated with the other digital forms of promotion and communication to become efficient (Tamaki, 2018).

There is an evident need to identify the most effective digital marketing techniques and tools in promoting a tourist destination, to attract and involve tourists (Castro *et al.*, 2017). A fundamental part of an organization's success in the digital age and the new information technologies is to be able to identify the needs of a client and to direct customized products and services to satisfy their desires (Buhalis and Law, 2008).

According to Uysal *et al.* (2011), websites are extremely important for tourist destinations, functioning as essential marketing tools, and they can be used to acquire huge amounts of data. For some destinations, the information on DMO websites is extensive, covering everything the tourist might need, with the marketing staff and website managers working together to collect and disseminate information that can be useful for the destination's promotion.

According to Kim and Fesenmaier (2008), tourists with well-defined goals have the habit of searching for the information they need through search engines and, as such, websites are a key part in this interaction and in this search for information. DMOs use their official websites to interact with tourists to promote the destination and provide relevant information about it. In this way, the website must be inserted in an efficient strategy where all other digital channels are oriented to make the tourist move there, and there must be regular updates of the existing information to always be able to satisfy the needs of a tourist who wants to have the best information possible.

3.3 Methodology

The objectives of the research carried out were to understand the main means of information used, both in the pre-trip phase and at the destination itself, identifying the opinion of tourists in relation to the digital technologies used by the DMO Tourism Center of Portugal. This study aimed to understand whether the digital channels must be the main form of interaction and communication and the main form of information dissemination for this and other destination management organizations, which are constantly confronted with change in a globalized world.

3.3.1 Questionnaire design

The methodology associated with the research question was realized using survey questionnaires directed to the tourists who visited Centre of Portugal. Check and Schutt (2012) consider that a survey based on questionnaires can be defined as the collection of information from a sample of individuals through their answers to questions. According to Pearce (2012), a survey based on questionnaires will have to involve certain considerations such as identifying the target population, selecting a representative sample, the correct design of the questionnaire and a clear definition of the administration method that it will use.

The questionnaire designed to answer this investigation was divided into two parts. The first part was aimed at understanding the different channels generally used by tourists in their pre-travel phase and at the destination itself, as well as the opinion of respondents about which channels should be used by the DMO to promote the destination they represent. The second part was designed to understand which channels were used in the pre-trip phase of the visit to Centre of Portugal and at the destination itself, to establish a direct comparison.

3.3.2 Sample definition

The number of questionnaires answered by tourists was 144. The questionnaires were delivered to the tourist offices of each location and concerned the different tourists who visited the spaces over a period of one month. The delimitation of the area of action of the research is an essential phase, trying to cover, as much as possible, the territory of Centre of Portugal, and forming a coherent response by the target population.

The different territories and the choice of municipalities where the questionnaires were administered were defined based on the analysis of the territories listed on the website of Centre of Portugal, which belongs to the Regional Tourism Promotion Agency of Centre of Portugal. In this way, 15 municipalities (Fig. 3.1) were identified in order to understand what would be the best strategy to determine the research population. The perception of the importance of these territories in Centre of Portugal proved to be necessary. Due to the lack of concrete data that would allow a clear count of the number of tourists from each of the municipalities, the number of overnight stays and the number of beds were defined as the statistics that were capable of describing the context of these territories in the general panorama of the Centre of Portugal region.

To understand the importance of these municipalities in the territory of Centre of Portugal, the number of overnight stays was considered. In summary, in 2018, Centre of Portugal recorded about 6,764,282 overnight stays, with these territories achieving a total of 4,361,609 overnight stays (INE, 2018). These municipalities represent about 64% of the total overnight stays in Centre of Portugal.

Fig. 3.1. Municipalities integrated in sample definition.

3.4 Tourism Marketing Organization in Portugal

3.4.1 Tourism of Portugal and Centre of Portugal

In Portugal, the tourism sector is under the responsibility of the Ministry of Economy and Digital Transition with a State Secretariat for Tourism, which is responsible for the development of tourism policies (Turismo de Portugal, n.d.a). The National Tourism Authority is called Turismo de Portugal, an organization integrated in the Ministry of Economy and Digital Transition, which has a national reach, being responsible for the promotion, valorization and sustainability of tourist activity, and comprising all institutional skills related to boosting tourism, both supply and demand (Turismo de Portugal, n.d.b).

Associated with this national tourism organization, there are five regional tourism entities, which have the mission of valuing and developing the tourist potential of the respective regional tourism area, as well as the integrated management of destinations within the framework of regional tourism development. These entities are designated in accordance with the NUT II where they are located (Turismo de Portugal, n.d.a).

These organizations are dedicated to the internal promotion of the regions where they operate, orienting their actions towards the enlarged internal market (national and cross-border with Spain). Oriented to the external regional promotion, with Regional Tourism Promotion Plans, there are Regional Tourism Promotion Agencies (RTPA) that, in conjunction with Tourism of Portugal, contribute to the execution of the National External Promotion Plan. According to Tourism of Portugal there are seven RTPAs considering the NUT II division of the territory.

The Centre Region (Fig. 3.2) is a NUT II region of Portugal, defined by DL no. 46/89. According to INE (2018), it has about 2,231,346 inhabitants. With Portugal having around 10,291,027 inhabitants, Centre of Portugal represents approximately 22% of the national population. The territory

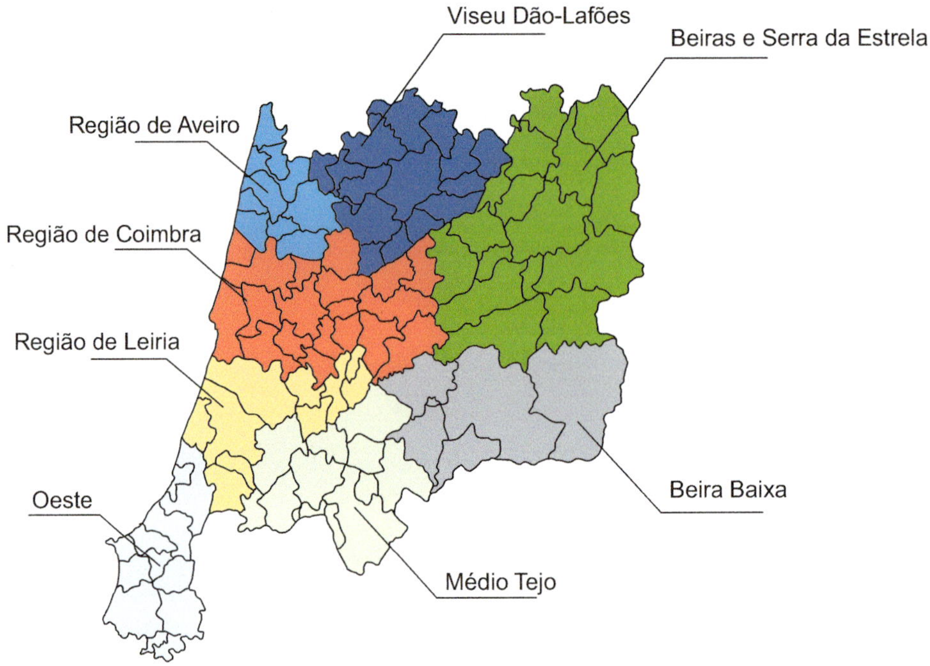

Fig. 3.2. Centre Region of Portugal.

encompasses 100 municipalities (Turismo Centro de Portugal, 2018) with a total area of 28,405 km^2, representing about 31.3% of the Portuguese territory, and is divided into 8 NUTs III (Pordata, n.d.).

The Centre Region stands out clearly from the rest of the country, with multiple differentiating dynamics in terms of political and geographical dimensions, being a region with different realities and discrepancies visible between the coast and what is inland, and being inserted between the two main drivers of national economic activities and two major metropolitan areas, Porto and the Lisbon metropolitan areas (Saur-Amaral *et al.*, 2013).

3.4.2 Regional organization of Tourism Center of Portugal

The Center of Portugal Regional Tourism Authority has as its mission:

> The enhancement and development of the tourist potential of the Centre Region, the internal promotion of the tourist destination, as well as the integrated management of the destination within the framework of regional tourism development. The management of Tourism Center of Portugal should be in accordance with the guidelines of the tourism policy defined by the Government and the multi-annual plans of the central administration and the municipalities that comprise it.
>
> (Office of the Secretary of State for Tourism, 2013, p. 21324)

Tourism Center of Portugal has an extremely relevant role in the development and promotion of tourist products in the region, focusing on the domestic market (Turismo Centro de Portugal, 2018).

Based in the city of Aveiro, Tourism Center of Portugal is the Portuguese tourism region that covers the largest territory, with around 100 municipalities and eight inter-municipal communities, integrating all the municipalities of Coimbra, Castelo Branco and Leiria, most of the municipalities of Viseu, Guarda and Aveiro and a third of the municipalities of Santarém and Lisbon (Turismo Centro de Portugal, 2018).

3.4.3 Characterization of the Regional Tourism Promotion Agency of Center of Portugal

According to Turismo de Portugal (n.d.c), the regional tourist promotion agencies, called RTPAs, are responsible for the external promotion of the respective regions, with the Regional Tourism Promotion Agency of Center of Portugal being responsible for the promotion of the Centre region. Regional tourist promotion agencies have two objectives: (i) tourism promotion of regions and their regional products through the study, preparation and development of specific actions in foreign markets; and (ii) co-ordination between regional public and private entities with a view to defining policies, objectives and strategies for the promotion of tourism in regions abroad. These objectives are oriented to the improvement of what is the tourist product of each of the regions, protecting the promotion of regional brands. In this way, the Regional Tourism Promotion Agency of Center of Portugal is oriented towards the promotion and dissemination of the Centre of Portugal region, while establishing public and private partnerships to achieve its objectives.

3.5 Analysis and Discussion of Results

3.5.1 Respondents' profiles

Through the observation of the questionnaires it is possible to present a profile of who is the tourist that visits Center of Portugal. Considering the 144 questionnaires answered, it is noticeable that around 83 (57.6%) were answered by females.

The analysis of the age groups of the respondents allows us to conclude that the tourists who visit the central region of Portugal are, in general, individuals over 45 years old (64%). The data on the level of education of the respondents allows us to conclude that tourists who travel to this specific region of Portugal are educated to degree level (76.6%).

A total of 34% of the respondents (49 tourists) stated that their country of residence was Portugal, with Spain and France ranking second and third, with 28 and 23 responses, respectively. About 21.5% of tourists (31) replied

Table 3.1. Means of obtaining information in the pre-trip phase.

Means of obtaining information	Uses	Doesn't use	Percentage of use
Website	70	74	48.6%
Social media	36	108	25%
Blog	18	126	12.5%
Brochures and leaflets	37	107	25.7%
Travel guides	44	100	30.6%
Travel agencies	17	127	11.8%
Digital forums	16	128	11.1%
Family and friends	41	103	28.5%
Books	11	133	7.6%
Mobile apps	20	124	13.9%
Reports and studies	3	141	2.1%
Tourism-related events	14	130	9.7%
Television and radio	9	135	6.3%

that they were from other countries that were not on the list presented, with only nine tourists indicating that their country of residence was Germany, and four from England.

3.5.2 Data analysis

The tourists were asked about the main means of obtaining information when choosing a destination, to understand if digital platforms were the main information sources used (Table 3.1).

Tourists were asked to choose only three of the available options, to establish which of them they used most regularly. Thus, analysing the responses (Table 3.1), it is possible to see that the first three options are websites, travel guides, and family and friends. It is possible to see that websites are, by some margin, the most frequent information source used, with travel guides and family and friends still quite important sources. Thus, tourists consider digital platforms their main means of obtaining information in their pre-trip phase.

Table 3.2. Main internet activity related to trips.

Activity	Uses	Doesn't use	Percentage of use
Searching for information about a destination	97	47	67.4%
Booking online	65	79	45.1%
Looking for inspiration	21	123	14.6%
Sharing tourist experiences	29	115	20.1%
Observation of reviews	22	122	15.3%
Buy or make a reservation to travel	28	116	19.4%
Search for accommodation at destination	47	97	32.6%

Table 3.3. Means of obtaining information at the chosen destination.

Means of obtaining information	Uses	Doesn't use	Percentage of use
Destination-owned websites	58	86	40.3%
Social networks of organizations belonging to the destination	26	118	18.1%
Blog	16	128	11.1%
Brochures and leaflets	49	95	34.0%
Travel guides	55	89	38.2%
Digital forums geared towards tourism and travel	8	136	5.6%
Family, friends and colleagues	22	122	15.3%
Books	13	131	9.0%
Mobile apps	22	122	15.3%
Tourist offices	63	81	43.8%
Destination reports and studies	10	134	6.9%

To ascertain whether the search for information is the most important activity of tourists on the internet, which proves the exceptional ability of the tool as a means of obtaining relevant data to support a trip, a question was inserted to help understand whether tourists identified the search for information as the main activity related to trips that they do on the internet.

The search for information (67.4%) is the main activity related to travel and tourism carried out by tourists on the internet, with the search for accommodation at the destination (32.6%) and online booking (45.1%) being the third and second most chosen options (Table 3.2). Tourists were also asked about their preferences in relation to the tools used to obtain information related to travel and tourism. To understand whether, at the destination itself, the choices reflected the importance of digital platforms in obtaining information, a question was added to the questionnaire.

While analysing the responses (Table 3.3), it is possible to observe that, in terms of obtaining information at the destination itself, the option 'Tourist offices' (43.8%) had the largest number of responses, followed by 'Websites of organizations belonging to the destination' (40.3%) and 'Travel guides' (38.2%). It is possible to conclude that tourists do not consider any of the digital platforms presented to be their main means of obtaining information at the destination; tourist offices assume that position, confirming them as the main source of information for travellers once at the destination.

3.5.3 Information sources used in the visit to Centre of Portugal

Analysing the visit to Centre of Portugal and the means of obtaining information used in the pre-trip phase (Table 3.4), it is possible to see that, where tourists were asked about the main means of obtaining information in a

Table 3.4. Means of obtaining information used in the Centre of Portugal pre-trip phase.

Means of obtaining information	Uses	Doesn't use	Percentage of use
Website	76	68	52.8%
Social media	30	114	20.8%
Blog	14	130	9.7%
Brochures and leaflets	53	91	36.8%
Travel guides	43	101	29.9%
Travel agencies	16	128	11.1%
Travel- and tourism-oriented digital forums	8	136	5.6%
Family, friends and co-workers	29	115	20.1%
Books	10	134	6.9%
Mobile apps	20	124	13.9%
Reports and studies	5	139	3.5%
Tourism-related events	11	133	7.6%
Television and radio	7	137	4.9%

more general way, the option 'Websites' (52.8%) was the most chosen by respondents, followed by 'Brochures and leaflets' (36.8%) and 'Travel guides' (29.9%). Once again, websites and travel guides are considered essential by tourists when looking for information, and, in the pre-trip phase of travelling to Centre of Portugal, brochures and leaflets also play an important role.

Tourists use digital platforms, such as websites, as the main means of obtaining information in the pre-trip phase of their trip to Centre of Portugal, with more than half of the respondents selecting this option in their choices. Looking at the specific context of the respondents' trips to Centre of Portugal (Table 3.5), it is possible to observe that the main activity carried out on the internet related to travel and tourism is the search for information about a destination (67.4%). The capacity of the internet as a means of obtaining information at the tourism level is confirmed, being presented by the respondents as the main one in a wide range of options.

Table 3.5. Main internet activity related to the Centre of Portugal trip.

Activities	Performs	Doesn't perform	Percentage of use
Searching for information about a destination	97	47	67.4%
Booking	60	84	41.7%
Looking for inspiration	17	127	11.8%
Sharing tourist experiences	25	119	17.4%
Observation of reviews	23	121	16.0%
Buy or make a reservation to travel	31	113	21.5%
Search for accommodation at destination	56	88	38.9%

Table 3.6. Means of obtaining information about Centre of Portugal.

Means of obtaining information	Uses	Doesn't use	Percentage of use
Destination-owned websites	59	85	41.0%
Social networks of organizations belonging to the destination	28	116	19.4%
Blog	14	130	9.7%
Brochures and leaflets	43	101	29.9%
Travel guides	34	110	23.6%
Digital forums geared towards tourism and travel	8	136	5.6%
Family, friends and colleagues	21	123	14.6%
Books	10	134	6.9%
Mobile apps	24	120	16.7%
Tourist offices	67	77	46.5%
Destination reports and studies	8	136	5.6%

It is possible to perceive that, when questioned in a broader context, tourist offices occupy a prominent position in obtaining information at the destination itself, and the websites of organizations belonging to the destination occupy the second position in the respondents' choices. Despite the prominence of websites, tourist offices obtained a greater number of responses (46.5%), having clear supremacy regarding the process of obtaining information upon arrival at the destination (Table 3.6).

Through the data analysis presented above it is possible to confirm that tourists use the internet and digital channels as their main form of information search in their pre-travel phase, and their visit to Centre of Portugal confirms this trend. The importance given by tourists to the promotion and communication of DMOs through digital channels is also clearly visible. Websites and social networks were the options most chosen by tourists when asked about the communication channels that should be used by organizations, followed by brochures and leaflets (Table 3.7). The option 'Websites' was chosen by 66% of respondents, which demonstrates the importance given to them as a communication tool (Table 3.7). As such, tourists believe that the main means of promoting a tourism region by destination management organizations is digital platforms.

It is therefore concluded that tourists consider the inclusion of digital technology by DMOs to be essential for communication with their target audience who consider that the means of communication that should be used by organizations that manage destinations is media inserted in a digital context, with emphasis on websites and social networks.

3.6 Implications and Conclusion

Clearly, technology is a hallmark of human success and is a fundamental part of society; it is woven into the daily lives of citizens in an organic way and it

Table 3.7. Means of promoting a tourism region by DMOs.

Promotion means	Considers it important	Doesn't consider it important	Percentage use
Social media	64	80	44.4%
Websites	95	49	66.0%
Blog	11	133	7.6%
E-mail	12	132	8.3%
Mobile apps	19	125	13.2%
Brochures and leaflets	61	83	42.4%
Attendance at events related to tourism	19	125	13.2%
Presence in digital forums oriented to the tourism and travel sector	13	131	9.0%
Advertising spots	14	130	9.7%
Presence in specialized press	4	140	2.8%
Destination reports and studies	12	132	8.3%
Creation of specialized guides	28	115	19.4%

is not possible to disassociate any element of everyday life from technological advancement. This has led to the world becoming extremely small, communication being almost instantaneous through a complex digital network, with the internet being a fundamental part of this globalization. Digital platforms, where websites and social networks are inserted, are changing not only the social concept but also the business concept. Everyone is impacted by technology, and everyone enjoys its benefits, enabling the creation of a complex and diversified network society where the central role belongs to technology and its dynamizing capacity.

Tourism, as an integral part of society and one of its most important sectors, has been affected by the technological phenomenon that we are experiencing. As it was possible to see in the literature review, tourism and technology are two of the greatest catalysts for enabling dynamic and innovative communities. Technology has become an integral part of the tourism sector and, for many authors, it is the 'heart' of the tourism industry, transforming the sector in terms of its capacity to transmit information and impacting tourists in all phases of their trip. In addition to the impact it brought to the tourist experience, technology and digital channels also allowed new dynamics to increase, with new solutions to be found to boost the sector, thus allowing it to evolve and continue to be a fundamental piece in this global chess game.

It is possible to see that the role of the DMO has changed. Its importance has increased and, at this moment, it is in a unique position to be the decisive factor in the development of a destination, with the aim of promoting it in a market that is highly competitive. It also influences the management of what a destination is, functioning as a factor of leadership. Thus it is possible to conclude that these organizations clearly have an extremely relevant role in the management and promotion of destinations, a major factor in the implementation of policies, and in the development and optimization of the tourist product and its promotion and marketing.

In Portugal, the tourism system is organized into five regional organizations, which have a series of specific missions, spread over the five NUTs II that exist in the country. For this study, the Centre of Portugal region was chosen as the main area of activity; Tourism Center of Portugal, a regional entity that has the mission of valuing and developing the tourist potential of the Centre of Portugal region, caters for the extended domestic market, and the Regional Tourism Promotion Agency of Center of Portugal for promotion to the external market.

Empirical research has given results that are clearly capable of being used to draw sustained conclusions. It was possible to see that the tourist who visits Centre of Portugal uses the internet and digital platforms such as websites and social networks in their search for information in the pre-travel phase, whereas at the destination there was a greater preference for tourist offices, followed by websites. Only 40% of the interviewed tourists stated that they used websites as an information source *at* the destination, with the remaining digital platforms having a very low usage rate, even with regard to the specific context of a trip to Centre of Portugal. This conclusion can also be explained by the somewhat advanced age of most of the respondents, which explains a certain traditionalism observed in the responses. However, it is an interesting conclusion that, despite a clear use of websites in the search for information at the destination, tourist offices are still considered a safe information source, more reliable than digital platforms which, except for websites, did not obtain high utilization rates.

It is therefore possible to conclude that tourists use digital channels mainly in their pre-trip phase, with tourism offices being the most important source of information at the destination.

Through the questionnaires, it was also possible to conclude that tourists see digital channels as the main promotion channels that should be used by DMOs to communicate with them, thus answering the question formulated at the developed methodology as to whether the digital channels must be their main form of interaction and communication and the main form of information dissemination.

The main implications associated with this study are naturally inserted in the theoretical review carried out, with the importance of digital platforms clearly highlighted, namely in the pre-travel phase. The relevance of websites is demonstrated in relation to other digital channels and cannot be neglected in terms of communication in Centre of Portugal. Assuming the theoretical revision carried out, with digital globalization and greater digital relevance in the tourism sector and in the various travel phases, which preponderance was identified in the research carried out, the communicational policies of the DMOs should be implemented through strong investment in digital communication, with updated and dynamic content aimed at answering any inquiries from tourists during the pre-trip phase, where the importance of digital channels is extremely relevant; neither should one neglect the communicational aspect at the destination itself which is capable of turning it into a secondary source of information. With the increasing introduction of technology in cities, it may become the main source of information used by tourists to orientate themselves in an unfamiliar environment.

The main limitation associated with this study was the inability to ensure responses across the entire territory associated with the Centre of Portugal region, given that the logistics would have been extremely complex. Future research could prove interesting if it had a greater geographic reach, enabling a global assessment of tourists visiting the national territory and drawing conclusions on a broader scale.

References

Adeola, O. and Evans, O. (2019) Digital tourism: mobile phones, internet and tourism in Africa. *Tourism Recreation Research* 27, 190–202.

Bitzinger, C. (2018) Authenticity – communication – culture: the role of ICTs and cultural collectives in the creation of authentic experiences in tourism destinations. In: Maurer, C. and Neuhofer, B. (eds) *ISCONTOUR 2018 – Tourism Research Perspectives: Proceedings of the International Student Confernece in Tourism Research*. Salzburg.

Buhalis, D. (2003) *eTourism – Information Technology for Strategic Tourism Management*. Pearson Education, London.

Buhalis, D. and Law, R. (2008) Progress in information technology and tourism management: 20 years on and 10 years after the Internet: the state of eTourism research. *Tourism Management* 29, 609–623.

Buhalis, D. and Zoge, M. (2007) The strategic impact of the internet on the tourism industry. In: Sigala, M., Mich, L. and Murphy, J. (eds) *Information and Communication Technologies in Tourism 2007*. Springer, Vienna, pp. 481–492.

Buhalis, D., Leung, D. and Law, R. (2011) eTourism: critical information and communication technologies for tourism destinations. In: Wang, Y. and Pizam, A. (eds) *Destination Marketing and Management: Theories and Applications*. CAB International, Wallingford, UK, pp. 205–224.

Castro, S.R., Silva, S.C.E. and Duarte, P. (2017) Does digital marketing really boost city tourism? Evidences from Porto's experience. *European Journal of Applied Business Management* 3, 84–100.

Chaffey, D., Ellis-Chadwick, F., Mayer, R. and Johnston, K. (2006) *Internet Marketing: Strategy, Implementation and Practice*, 3rd edn. Prentice Hall/Financial Times, Englewood Cliffs, New Jersey.

Check, J. and Schutt, R.K. (2012) *Research Methods in Education*. SAGE Publications, Thousand Oaks, California.

Da Silva Pina, A.J. (2015) *La usabilidad del sitio web de un destino turístico y sus efectos en la demanda: una aproximación a un índice de satisfacción digital turística*. Universidad de Extremadura.

Da Silva Pina, A.J., Hernández-Mogollón, J.M. and Campon-Cerró. (2017) A usabilidade de um website de um destino turístico e os seus efeitos na procura : aproximação a um índice de satisfação digital. *Journal of Tourism & Development* 27/28, 733–744.

Gretzel, U., Sigala, M., Xiang, Z. and Koo, C. (2015) Smart tourism: foundations and developments. *Electronic Markets* 25, 179–188.

INE (2018) População residente (N.o) por Local de residência (NUTS – 2013). Available at: https://www.ine.pt/xportal/xmain?xpid=INE&xpgid=ine_indicadores&contecto=pi&indOcorrCod=0008273&selTab=tab0&xlang=pt (accessed 13 April 2019).

Kannan, P.K. and Li, H.A. (2017) Digital marketing: a framework, review and research agenda. *International Journal of Research in Marketing* 34, 22–45.

Kaur, G. (2017) The importance of digital marketing in the tourism. *International Journal of Research Granthaalayah* 5, 72–77.

Kim, H. and Fesenmaier, D.R. (2008) Persuasive design of destination web sites: an analysis of first impression. *Journal of Travel Research* 47, 3–13.

Kotler, P., Keller, K.L., Brady, M., Goodman, M. and Hansen, T. (2009) *Marketing Management*. Pearson Education Limited, Harlow, UK.

Law, R., Chan, I.C.C. and Wang, L. (2018) A comprehensive review of mobile technology use in hospitality and tourism. *Journal of Hospitality Marketing & Management* 27, 626–648.

Leung, D., Law, R., van Hoof, H. and Buhalis, D. (2013) Social media in tourism and hospitality: a literature review. *Journal of Travel and Tourism Marketing* 30, 3–22.

Li, S., Robinson, P. and Oriade, A. (2017) Destination marketing: the use of technology since the millennium. *Journal of Destination Marketing and Management* 6, 95–102.

Madarasz, E. and Zsofia, P. (2011) Using network analysis to define a tourist destination as the basis for a DMO. *International Conference on Tourism and Management Studies,* 1036–1039.

Mariani, M., Buhalis, D., Longhi, C. and Vitouladiti, O. (2014) Managing change in tourism destinations: key issues and current trends. *Journal of Destination Marketing and Management* 2, 269–272.

McFee, A., Mayrhofer, T., Bar tov , A., Neuhofer, B., Rainoldi, M. *et al.* (2019) The effects of virtual reality on destination image formation. In: Pesonen, J. and Neidhardt, J. (eds) *Information and Communication Technologies in Tourism 2019*. Springer, Cham, Switzerland, pp. 107–119.

Minguzzi, A. (2006) Destination competitiveness and the role of Destination Management Organizations (DMO): an Italian experience. In: Lazzeretti, L. and Petrillo, C.S. (eds) *Tourism Local Systems and Networking*. Elsevier.

Morosan, C. (2008) DMO websites and the role of complementary media in tourism advertising. *Journal of Hospitality and Leisure Marketing* 17, 216–236.

Musina, Ž. and Gao, X. (2016) DMO tourism website's success evaluation model and framework. *European Journal of Economics and Management Sciences* 16, 61–76.

Office of the Secretary of State for Tourism (2013) Order no. 8864/2013: Official Gazette no. 129/2013, Series II of 2013-07-08. Series II Act Ministry of Economy and Employment. Approval of the Statutes of the Regional Tourism Entities of the Center of Portugal, Alentejo and Algarve.

Oliveira, E. and Panyik, E. (2014) Content, context and co-creation. *Journal of Vacation Marketing* 21, 53–74.

Ong, Y.X. and Ito, N. (2019) I want to go there too! Evaluating social media influencer marketing effectiveness: a case study of Hokkaido's DMO. In: Pesonen, J. and Neidhardt, J. (eds) *Information and Communication Technologies in Tourism 2019*. Springer, Cham, Switzerland, pp. 132–144.

Pearce, D. (2012) *Frameworks for Tourism Research*. CABI International, Wallingford, UK.

Pike, S. and Page, S.J. (2014) Destination Marketing Organizations and destination marketing: a narrative analysis of the literature. *Tourism Management* 41, 202–227.

Pitana, I.G. and Pitanatri, P.D.S. (2016) *Digital marketing in tourism: the more global, the more personal*. Promoting Culture and Heritage Culture, International Tourism Conference, pp. 116–125.

Pomirleanu, N., Schibrowsky, J.A., Peltier, J. and Nill, A. (2013) A review of internet marketing research over the past 20 years and future research direction. *Journal of Research in Interactive Marketing* 7, 166–181.

Pordata (n.d.) O que são NUTS? Available at: https://www.pordata.pt/O+que+sao+NUTS (accessed 13 April 2019).

Prahalad, C.K. and Ramaswamy, V. (2004) Co-creation experiences: the next practice in value creation. *Journal of Interactive Marketing* 18, 5–14.

Presenza, A., Sheehan, L. and Ritchie, J.R.B. (2005) Towards a model of the roles and activities of destination management organisations. *Journal of Hospitality, Tourism and Leisure Science*.

Qi, S., Law, R. and Buhalis, D. (2008) Usability of Chinese destination management organization websites. *Journal of Travel and Tourism Marketing* 25, 182–198.

Ritchie, J.R.B. and Crouch, G.I. (2003) *The Competitive Destination – A Sustainable Tourism Perspective*. CAB International, Wallingford, UK.

Saur-Amaral, I., Aragonez, T., Gouveia, M., Damas, D. and Costantino, F. (2013) *Plano de Marketing da Turismo Centro de Portugal – Parte I: Análise Interna e Externa*. IPAM Lab.

Schmidt, S., Cantallops, A.S. and dos Santos, C.P. (2008) The characteristics of hotel websites and their implications for website effectiveness. *International Journal of Hospitality Management* 27, 504–516.

Sheehan, L., Vargas-Sánchez, A., Presenza, A. and Abbate, T. (2016) The use of intelligence in tourism destination management: an emerging role for DMOs. *International Journal of Tourism Research* 18(6), 549–557. DOI: 10.1002/jtr.2072.

Smith, K.T. (2011) Digital marketing strategies that Millennials find appealing, motivating, or just annoying. *Journal of Strategic Marketing* 19, 489–499.

Sotiriadis, M.D. (2017) Sharing tourism experiences in social media. *International Journal of Contemporary Hospitality Management* 29, 179–225.

Straker, K. and Wrigley, C. (2016) Translating emotional insights into digital channel designs. *Journal of Hospitality and Tourism Technology* 7, 135–157.

Tamaki, K. (2018) Integrated, diverse media marketing linked with database marketing for inbound reginal tourism planned by destination. *Journal of Global Tourism Research* 3, 43–48.

Tanrisevdi, A. and Duran, N. (2011) Comparative evaluation of the official destination websites from the perspective of customers. *Journal of Hospitality Marketing and Management* 20, 740–765.

Turismo Centro de Portugal (2018) *Plano de Atividades e Orçamento 2019*.

Turismo de Portugal (n.d.a) Organização e Parceiros. Available at: http://www.turismodeportugal.pt/pt/Turismo_Portugal/Organizacao_Parceiros/Paginas/default.aspx (accessed 17 February 2019).

Turismo de Portugal (n.d.b) Turismo de Portugal: Quem somos e o que fazemos. Available at: http://www.turismodeportugal.pt/pt/quem_somos/Organizacao/information-hub-mne/Paginas/turismo-de-portugal-quem-somos-e-o-que-fazemos.aspx (accessed 18 February 2019).

Turismo de Portugal (n.d.c) Agências Regionais de Promoção Turística. Available at: http://business.turismodeportugal.pt/pt/Conhecer/Quem_e_quem/Paginas/Agencias-Regionais-de-Promocao-Turistica.aspx (accessed 19 February 2019).

Tussyadiah, I.P. and Fesenmaier, D.R. (2009) Mediating tourist experiences. access to places via shared videos. *Annals of Tourism Research* 36, 24–40.

Updhya, A. (2014) Role of Destination Management Organization in tourism crisis management: a Middle Eastern perspective. *3rd Interdisciplinary Tourism Research Conference*, Istanbul. DOI: 10.13140/RG.2.1.3881.6803.

Uysal, M., Harrill, R. and Woo, E. (2011) Destination marketing research: issues and challenges. In: Wang, Y. and Pizam, A. (eds) *Destination Marketing and Management: Theories and Applications*. CAB International, Wallingford, UK, pp. 99–112.

Vyas, C. (2019) Evaluating state tourism websites using Search Engine Optimization tools. *Tourism Management* 73, 64–70.

Werthner, H. and Ricci, F. (2004) E-commerce and tourism. Communications of the ACM. Available at: https://doi.org/10.1145/1035134.1035141 (accessed 11 June 2021).

Wymbs, C. (2011) Digital marketing: the time for a new 'academic major' has arrived. *Journal of Marketing Education* 33, 93–106.

Yang, X. and Wang, D. (2015) The exploration of social media marketing strategies of destination marketing organizations in China. *Journal of China Tourism Research* 11, 166–185.

4 New Tourism Experiences and Mature Destinations: The Role of Participatory Planning in Repositioning a Southern Italy Destination

Francesco Calza, Michele Simoni, Annarita Sorrentino* and Mariapina Trunfio

Parthenope University of Naples, Italy

4.1 Introduction

The restrictions associated with the Covid-19 pandemic have emphasized the paradigms already evident before the pandemic, in which tourists were increasingly oriented towards the search for unique, authentic, tailor-made and memorable experiences. With the limitations imposed by the pandemic and the inability to travel, as well as factors such as life at home, an intense relationship with digital devices – both for working and for staying in touch with loved ones – greater attention to health and the environment, and the desire to enjoy vacations to improve personal well-being and for psycho-physical and cultural enrichment, is increasing more than ever before.

Today, more than before, tourists are becoming increasingly demanding in their travel experience. They want to perceive security, authenticity, protection, mindfulness, learning (Holbrook and Hirschman, 1982; Mehmetoglu and Engen, 2011; Miao *et al.*, 2021), and to have positive memories about their experience to share with friends and relatives. Coherently, it seems reasonable to expect that destinations move beyond providing just accommodation, transportation and food to engaging people in an experiential visit.

Postmodernism had already outlined a scenario where consumers were increasingly segmented (Lew, 2008; Egger *et al.*, 2020) and increasingly looking for a hedonic motivation (e.g. escape, personal growth, well-being and authenticity) more than a destination visit (Ejarque, 2003). With the pandemic

*Corresponding author: annarita.sorrentino@uniparthenope.it

DOI: 10.1079/9781789248753.0004

crisis, the stay-at-home necessity and social distancing, there is a need for destinations to reinvent their value propositions by designing their offering and communications around experience and quality (Bazzanella *et al.*, 2020), guaranteeing safety and experiential value and finding new ways to generate value and distribute income in the community.

Against this backdrop, it is becoming even more important for destinations to focus on core businesses, eliminate excess supply and rethink new tourist offers based on authentic experiences (Bazzanella *et al.*, 2020) by eliminating standardized and generalist products and designing differentiated and experience-based products. Indeed, the global health emergency has prompted destinations to reflect on the importance of environmental sustainability and, in general, on the attention to primary needs as well as the need to rethink business models. Although everything is still constantly changing and there are no answers, it is evident that, at least in the short term and since there will be fewer tourists, destinations need to focus on taking care of the welcome in the smallest detail, guaranteeing safety and experiential value.

The experiential perspective seems to be one way to attract new segments and retain existing segments (Schmitt, 2010; Ferraresi and Schmitt, 2018). However, this perspective is only part of a broader strategic model that mainly involves community-type destinations in which the tourist offer is entrusted to a plurality of stakeholders who, through various and variable forms of interaction, can manage attraction factors and create joint value for the destination, and, consequently, unique and unreproducible tourist experiences (Pechlaner *et al.*, 2010; Lucia and Buffa, 2011). This approach requires destinations to differentiate themselves through a relationship-based strategy involving all stakeholders within the local community.

This approach is more difficult to pursue in those destinations where there is neither formalized governance nor a shared tourism strategy. This is the case with 'demand-driven' destinations where only natural resources have been enough to attract tourists, and the tourism development processes have been driven by spontaneous demand for years rather than a formal destination management strategy. However, in a scenario as difficult and complex as the one outlined thus far, it is not possible to prescind from a sharing of common development policies to maintain competitiveness. The development of a sharing of goals and actions to attract and manage the tourist demand becomes the first step for the definition of a destination identity and a valuable destination experience (Ruhanen, 2009; Pyke *et al.*, 2018; Trunfio and Della Lucia, 2019).

This chapter aims to define a repositioning model for demand-driven destinations starting from a participatory planning process involving local stakeholders. This model was applied to the case of Centola-Palinuro, a seaside destination located in the Campania region of southern Italy that is considered to be demand-driven due to its spontaneous tourist flows, but suffering from the limits of seaside seasonality, mass-market demand and low spending. Results show four possible strategic directions for the development of a shared destination management model by enriching existing destination management literature with an empirical case and offering practical suggestions for policymakers and destination managers. In the conclusion, the authors argue that stakeholder engagement is crucial for capturing the interest of local players and their loyalty to build a cohesive group of people aligned to achieve shared goals.

4.2 Theoretical Background

4.2.1 Understanding demand trends to develop new tourism experiences

The evolution of paradigms in tourism, from traditional to postmodern tourism, can be analysed by considering three significant aspects: the social function of holidays, external attitudes towards tourism, and target models (Tribe *et al.*, 2015). What emerges is that postmodern tourism is characterized by some changes, from tourism as a function of relaxation and recreation, which is typical of mass-market tourist products, to the desire to travel to meet similar subjects who have the same lifestyles, similar interests and hobbies, and similar ways of enjoying themselves. The holiday is no longer an indicator of social status but a tool for personal and emotional satisfaction in which authentic and co-produced experiences are sought. The segmentation of tourism demand shifts from traditional socio-demographic segmentation models, which are no longer able to describe the different attitudes and purchasing and consumption behaviours of tourists (Bigné and Andreu, 2004; Bigné *et al.*, 2008; Millán *et al.*, 2016; Risitano *et al.*, 2020), to perceptual criteria based on the emotions sought.

For example, in cruise tourism, Risitano *et al.* (2020) have profiled passengers based on the experience response to the onboard atmosphere by revealing that cruise passengers who perceive the cruise vacation to be a passive experience are active women who are not willing to share their experience through word-of-mouth (WOM), while men are more likely to share their experience with others.

These changes have led to substantial changes in the psychographic profiles of tourists, which are numerous, articulated and complex compared to the profiles of the last century, and in the reasons for the trip, which significantly impact the traveller's decision-making process through the various stages of the journey (acquisition of information, evaluation of alternatives in terms of the destination and facilities/services, choice, purchase, experience, word-of-mouth, return) (Yachin, 2018; Shen *et al.*, 2020).

Several studies (Moscardo, 2011; Mahika, 2011; Bizirgianni and Dionysopoulou, 2013; Richards, 2015; Buffa, 2015; Yousaf *et al.*, 2018) highlight the change in the tourist profile and the tourist's travel motivations. Indeed, digitalization and low-cost airlines (Ejarque, 2003) have questioned the entire decision-making of tourists who in recent years have become increasingly autonomous in the design of their holiday and demanding in what they are looking for; that is, getting away from the daily routine and feeling, temporarily, a resident in an authentic place (Egger *et al.*,

2020). This change undermines the traditional models of tourism offers and requires radical innovations in the relationships of tourism demand, tourism companies and destinations. For example, it is increasingly necessary to intercept demand to stimulate the emotions sought by travellers from the pre-purchase phase (e.g. escape, education, mindfulness) with captive and immersive communication. It is important to ensure the right relationship between efficient services and friendliness of the local community, and that destinations and tourism players are aligned on common goals.

Even before the pandemic, some major changes in tourist behaviours occurred, from a concept of passive and functional travel, typical of the mass-market tourist products of the 1980s and 1990s, to the desire to travel for forms of special interest tourism: meeting similar subjects who have similar lifestyles, interests and hobbies and similar ways of enjoying the holiday (Ejarque, 2009; Martini, 2017; Presenza and Formato, 2018). Therefore, the journey becomes an opportunity to co-create value by becoming sustainable and responsible because it responds to the desire to relate to local cultures, to preserve the environment, to become a temporary resident of places and to live slowly. Thus, the new travel/vacation configuration presupposes an orientation towards the traveller rather than towards the offer, both by individual companies and at the destination level. This orientation must include escape experiences that are rich in emotional and learning value and modelled in a customized way for each traveller (Calza *et al.*, 2019a).

In this scenario, a hyper-segmentation emerges that highlights new targets in search of immaterial value rather than value for money, which still represents a motivation for choice but is residual and applies only for some customer categories, for example millennials (Yousaf *et al.*, 2018; Egger *et al.*, 2020). It also highlights the search by tourists for an increasingly slow holiday in contact with nature but, at the same time, 'active', where there is room for the active participation of the person in the relationship with the context in which he or she is staying. In this regard, a form of botanical tourism is emerging, in which simple contact with nature becomes something that stimulates the senses and makes the travel experience of great sensory and emotional impact, leading to specialization on the part of operators.

The same can be said for the concept of 'wellness tourism', which is currently understood in a much broader sense than the classic 'spa' concept. Well-being has become, above all, mental; therefore, forms of vacation linked to detox, yoga stays, running and mindfulness are consolidating. The so-called tourism of mobile photography is also gaining momentum. In this form of tourism, beauty, exclusivity and authenticity are sought to capture emotions to be transmitted and shared through social media. Therefore, a profile emerges of a traveller who seeks hedonistic value and aims at the satisfaction of personal values as well as an authentic form of relationship that seeks contact with nature and with locals.

From the analysis of the reasons set out thus far, something resembling a dichotomization emerges between reasons that confirm consolidated forms of tourism, such as beach, food and wine, lakesides and mountains, and emerging forms of tourism that profile new tourists and new offers, such as tourism linked to mindfulness, sensory contact with nature (botanical tourism), mobile photography or 'selfie-gaze tourism' and, more generally, a form of tourism that can be defined as relational, that is tourism that puts the guest in contact with locals in the community and with the territory in all its forms and that ultimately is transversal to all forms of motivation (Pesonen and Komppula, 2010; Smith and Diekmann, 2017). These are current trends that emerged from the recent studies (Dogru *et al.*, 2017; Egger *et al.*, 2020; da Mota and Pickering, 2020) where some themes are substantially updated with forms of travel closer to the identity of today's tourist (Rittichainuwat, 2018). After reviewing the main literature on tourism motivations and considering the current global situation characterized by a limitation in travel, home-working and digital relationships, we consider current demand trends to be mindfulness tourism (Moscardo, 2009; Stankov *et al.*, 2020), digital detox tourism (Hoving, 2017) and selfie-gaze tourism (Sigala, 2019).

Concerning mindfulness tourism, this kind of trend is related to a new concept of well-being, both physical and mental, and subjectively perceived by each individual (Frauman and Norman, 2004; Moscardo, 2011; Stankov *et al.*, 2020). In other words, this kind of well-being is defined as mindfulness, which is a mental process that contributes to fighting emotional distress and maladaptive behaviour (Bishop *et al.*, 2004, p. 230; Stankov *et al.*, 2020, p. 704). In the tourism field, it has been recognized for its beneficial effects on tourist well-being and its transformative influences on tourist experiences (Chen *et al.*, 2017; Kirwin *et al.*, 2019; Loureiro *et al.*, 2019). The concept of wellness has thus far been associated with a moment of passive leisure related to bodily relaxation (Stará and Peterson, 2017). However, recently, and especially during difficult times after the pandemic crisis, people have been looking for moments to devote to reflection and spiritual well-being. Against this backdrop, well-being becomes reflection, awareness, meditation and silence, and it becomes mindfulness.

The second trend related to the concept *of exclusivity* is linked to the new concept of luxury. From traditionally being linked to the concepts of high quality standards, customization and high price, it has a new meaning in which the role of the hedonistic value of luxury is emphasized and linked to refinement, rarity and silence. The personalization of services and the culture of hospitality synthesize exclusivity (Batat, 2019).

Thirdly, travel attitudes are moving towards an exchange of meanings, no longer a mere exchange of services and the new forms of well-being tend to be increasingly 'active'; that is, they require the active participation of the tourist and immersive involvement in the territory. Sports emerge as primary travel motivations, but above all they need to feel active during the holiday, for example by participating in culinary tours, birdwatching or walks aimed at recognizing wild herbs and then participating in their preparation before a tasting.

Another paradigm linked to the new concept of luxury that emerges is the concept of digital detox (Carciofi, 2017). In a society where people find themselves spending their days overloaded with information and always virtually connected, the value of time becomes lost. Thus, detox trips are being created. In such trips, one discovers and rediscovers the value of connecting with oneself and with others; that is, the value of luxury.

Finally, we observed how the need to take a vacation is emerging in a latent way to fuel one's recognition and popularity in the world of virtual connections. This is the so-called mobile photography tourism (or selfie-gaze tourism). Travellers reach destinations so that they may photograph themselves in unique places with the sole purpose of sharing these images with their virtual networks (Sigala, 2019; Egger *et al.*, 2020).

4.2.2 Engaging stakeholders with participatory planning processes

To capture demand, a territorial offer must be as cohesive and structured as possible. However, it is often difficult to find destinations where the different players work to pursue a collective goal because doing so requires governance that can set long-term strategic objectives and co-ordinate local stakeholders. Although academics and destination managers have widely recognized the importance of stakeholder engagement in the development of competitive destinations, considerable challenges remain (Trunfio and Della Lucia, 2019). In virtuous destinations, destination management organizations (DMOs) play a significant role in local stakeholder engagement by reconciling the interests of, requests from and benefits for multiple-destination stakeholders. Various networks of regional institutions and public and private stakeholders contribute to setting institutional mechanisms and/or (in)formal processes to legitimize political power, share authority and responsibility for destination decision-making, balance interests and resolve possible conflicts (Atorough and Martin, 2012; March and Wilkinson, 2009; Quinlan *et al.*, 2013).

The destination management literature shows that socioeconomic change has prompted an evolution towards shared governance models. Both 'collaborative' (Ansell and Gash, 2008) and 'embedded' governance' (Go and Trunfio, 2012) relate to participatory decision-making (Ruhanen *et al.*, 2010; Choibamroong *et al.*, 2011; Bramwell and Lane, 2011; Beritelli, 2011; Go *et al.*, 2013), leading to destination value co-creation (Cabiddu *et al.*, 2013). To socialize the 'many' in the decision-making process, DMOs or policymakers must identify tools for supporting and managing destination stakeholder engagement. These tools may be weak (information, consultation) or strong (participation, co-operation) (Plaza-Úbeda *et al.*, 2010). Participatory tools include public hearings, workshops, advisory committees, social contact events, negotiations and public engagement (e.g. surveys, focus groups, invitations to submit written comments) (Byrd, 2007).

According to Quinlan *et al.* (2013, p. 10), 'Stakeholder engagement is…a philosophy, a strategy, an organisational capability, a process and a range of interaction instruments' that create 'opportunities for innovation and growth', and it includes three steps: attraction, integration and management. The attraction is the first stage of the engagement process. It aims to build stakeholder awareness and interest in participation. Integration is the stage for interactive networking aimed at developing positive relationships, thereby enabling the pursuit of shared objectives. The management stage establishes activities of monitoring and motivation that enhance practice and optimize collaborative outcomes (Waligo *et al.*, 2013). The number and types of stakeholders involved can both determine and limit the participation instruments utilized (Bramwell and Sharman, 1999). In the present research, we used this framework as a first attempt to experiment with stakeholder engagement in a small coastal destination that wants to create participatory governance to better organize local tourism management. Particularly, in this first exploratory stage, we implemented the 'attraction' stage (Quinlan *et al.*, 2013) through thematic meetings named 'thematic tables' with several local stakeholders as a type of participatory planning process.

4.3 Methodology

To achieve the aim of this research, we use an exploratory case study approach (Yin, 2017) and study the case of Centola-Palinuro, which is a small coastal mass-market tourism destination in southern Italy in the Campania region. The case study method is useful for an exploratory study that 'focuses on understanding the dynamics present within single settings' (Eisenhardt, 1989, p. 534).

The analysis was carried out in two phases aimed at exploring the phenomenon from the supply side and the demand side. In the first phase, desk research was conducted on two levels: scenario research on the main emerging trends at the international level and research on the territory to map the trends of local tourist demand and the composition of the offer. In the second phase, a field analysis was conducted through thematic meetings with local stakeholders, which allowed us to trigger the participatory planning process, starting with the involvement of all categories of local actors (e.g. hotels, camping sites, restaurants, bed & breakfasts, policymakers), which is defined in the literature as stakeholder engagement (Trunfio and Della Lucia, 2019). In particular, four thematic meetings were organized on four different topics: I. tourist experiences; II. quality management; III. accessibility, hospitality and information; and IV. innovation and training.

The participants were selected according to their local representativeness as tourism stakeholders (e.g. hotels, bed & breakfast, restaurants, excursion agencies) through a list prepared by the research team and the Municipality of Centola. Afterwards, hard-copy formal invitations (or e-mail in some cases) followed by an initial contact by telephone were used as an invitation. A total of 50 tourism leaders from the private and public sector agreed to participate in the discussions.

An average of 50 people attended each meeting. These participants represented hotels owners (50%), bed & breakfasts (10%), restaurants (20%), incoming agencies (10%), local associations and small tourism businesses such as bike and boat rental businesses and excursion organizations (10%). The meetings were held at the council chamber of the Municipality of Centola one afternoon per week during the 2019 winter season.

The meetings were guided by the three researchers, who presented the topic of the day and initially stimulated discussion and interaction with some opening questions. During the discussion, two researchers moderated the discussion, while the third took notes on the strengths and weaknesses that emerged. Each meeting lasted four hours. Data and information were transcribed and analysed manually in this stage to recognize and map all the interventions. Transcripts were analysed by using the coding process of dividing data into codes and categories related to the segments (Miles and Huberman, 1994). Interpretation of data collected helped in informing and further explaining the study results.

4.3.1 The case of Centola-Palinuro: background and setting

The destination of Centola-Palinuro is a small coastal town in southern Italy (Campania region) characterized by a historical and undisputed seaside vocation. It has a history with historical-mythological elements that represent an important reference in the possible processes of valorizing the same historical-mythological memory for tourism purposes. From a seaside perspective, with its several beaches and cases, the primary vocation of Centola-Palinuro is an expression of unique and distinctive natural resources that currently represent the main factors of attraction. The distinctiveness of these resources has allowed the destination to receive some prestigious quality certifications including, in 2018, the recognition of the Italian Blue Flag, which is a prestigious recognition that certifies the suitability of the waters for swimming. Finally, in 2019, the Municipality of Centola joined the Wine Cities network to embark on a path to enhance public land with new vineyards. Its predominantly seaside vocation is one of the reasons for the concentration of demand in the summer period, which highlights a strong tourist seasonality with peaks in attendance in June, July and August (see Table 4.1). From a demand-side perspective, incoming tourism is characterized by the national and local demand of families. It is a traditional coastal mass-market destination where families spend their main holiday during the summer. From a supply-side perspective, the accommodation offerings consist of almost 160 structures with a total of 6815 beds, but holiday homes have become a major phenomenon. Tourism service activities are provided by almost 80 small organizations, including restaurants, bars and entertainment service enterprises, such as canoe rental, pedal-boat rental and windsurfing training businesses (Calza *et al.*, 2019b). From a destination management perspective, this destination suffers from the phenomenon of overtourism during the summer months from June to August. From September, all activities shut down until the next season, resulting in a so-called 'ghost city' effect. From a managerial perspective, Centola-Palinuro does not have formal governance (e.g. the presence of a DMO) or a shared management model. However, local stakeholders would like to be guided towards a more organized and profitable flow-management model. This profile reveals a destination that can be considered demand-driven due to its spontaneous renown in attracting

Table 4.1. Arrivals and room nights. (Authors' elaboration from official data of local tourism officer, 2020.)

	2018		2019	
	Arrivals	Room nights	Arrivals	Room nights
January				
February				
March	33	135		
April	94	323	105	228
May	205	644	213	794
June	1343	11,541	3422	10,627
July	1136	6920	1228	7568
August	4067	14,979	5535	15,650
September	525	2891	175	719
October	18	67	45	104
November				
December				

tourist flows but suffering from the limitations of seaside seasonality, mass-market demand and low spending.

4.4 Results

The four thematic meetings constituted the first significant result in developing a shared development tourism plan for Centola-Palinuro involving 50 local actors who concretely contributed to defining the strengths and weaknesses of the destination. The significant participation of local actors and the high representation of the various categories (e.g. hotel and extra-hotel accommodation facilities, tourism businesses related to various activities in the area, trade associations, tour operators, citizens) made it possible to reach a shared definition for each table of strengths, weaknesses, possible solutions and priority objectives for the future. The weaknesses and strengths that emerged from the thematic meetings are summarized in Tables 4.2 and 4.3, respectively. Notably, the participants at the meetings highlighted aspects considered particularly significant for the future development of Centola-Palinuro, even if not strictly linked to the thematic meeting. As shown, there are issues and proposed solutions that constantly recur in the various thematic meetings, testifying to the significance of some aspects perceived by the actors.

As noted in Table 4.2, Centola-Palinuro is known as a seaside destination, which is the core destination vocation, and it is characterized by a mass-market tourism model aimed at families, typical of the 1980s–1990s. The countryside and its agricultural heritage are almost unknown.

Table 4.2. Weaknesses that emerged from the thematic tables.

Critical issues	I. Tourist experiences	II. Quality management	III. Accessibility, hospitality and information	IV. Innovation and training
Destination has lost its notoriety and tourist attractions over the years.	✓	✓		✓
Tourism has developed spontaneously thanks to the natural beauty (sea and nature) and the marketing actions of some excellent accommodation facilities.	✓	✓	✓	✓
Tourism is seasonal and concentrated in two months of July and August in the Palinuro area.	✓	✓	✓	✓
Overtourism negatively affects the quality of all destination services, in particular on mobility, parking and beaches.		✓	✓	
'Ghost city' effect during autumn and winter seasons.			✓	✓
Presence of two territories with different degrees of tourism development: the coast, with established forms of seaside tourism and mass-market; the countryside with high potential for the development of alternative and niche forms of tourism (naturalistic, sports, food and wine), but with spontaneous and unstructured forms of tourism.	✓	✓	✓	
Predominant demand segment of the bathing establishment is the family with regional origin (about 50-60%) and with a marginal weight of foreign European tourists.	✓	✓		
Urban environment needs care, including by citizens, and is one of the priorities for development and especially for the repositioning of the destination.	✓		✓	
Natural environment is often degraded and/or poorly cared for and needs regulation and controls.			✓	✓
Need to improve accessibility, mobility and parking.		✓	✓	

Continued

Table 4.2. Continued

Critical issues	I. Tourist experiences	II. Quality management	III. Accessibility, hospitality and information	IV. Innovation and training
Presence of illegal forms and non-regulated activities (e.g. second homes).		✓		
Absence of dialogue, collaboration and integration between the different actors/tourist offers.				✓
Absence of an integrated communication and tourist information system.	✓	✓		
Need to sensitize tour operators to the extension of the season (at the moment several accommodation facilities are also open in autumn and spring).			✓	✓
Low entrepreneurship (including young people) related to new forms of tourism and technologies.	✓	✓		✓
Many tourism players require managerial and professional training (e.g. improvement of English skills, hospitality and service management training).			✓	✓

Table 4.3. Strengths that emerged from the thematic tables.

Strengths	I.Tourist experiences	II. Quality management	III. Accessibility, hospitality and information	IV. Innovation and training
Uncontaminated and certified natural heritage (Italian blue flag and 5 Vele-Legambiente).	✓	✓	✓	
Image and reputation of Cilento.	✓	✓	✓	✓
Organizations and associations that promote forms of tourism that make it possible to differentiate the traditional seasonal product (sun, sea and sand): underwater activities with the presence of groups of biologists who study the seabed; water sports such as canoeing, sailing, skiing, kayaking, norkeling, underwater trekking, daily cruises, excursions to caves, sailing school, pleasure boating, fishing tourism.	✓	✓	✓	✓
Organizations and associations that promote forms of supplementary and/or alternative tourism to the seaside: trekking; botanical tourism; primrose festival, etc.	✓	✓	✓	✓
Projects to enhance the countryside, in particular the village of Sanseverino di Centola. Twinning with a city in the USA to create the Museum-House of the Emigrant and the Park of Memory.	✓	✓	✓	✓
Quality extra-hotel accommodation and related to the recovery of the identity and tradition of the places.	✓	✓	✓	✓
A varied food and wine offer and the candidacy as a city of wine.	✓	✓	✓	✓

In addition to the various weaknesses of the tourist destination, the thematic meetings revealed the abundant potential for the development and differentiation of tourist experiences throughout the destination. The weaknesses that emerged from all tables have been: 'Tourism has developed spontaneously thanks to the natural beauty (sea and nature) and the marketing actions of some excellent accommodation facilities' and 'Tourism is seasonal and concentrated in two months of July and August in the Palinuro area'. That means that local players are not satisfied with the concentrated overtourism during summer and the lack of local governance that can intercept a different type of demand rather than that which spontaneously reaches the destination. This result is confirmed by another topic that emerged from the majority of the meetings, that is,

> 'Presence of two territories with different degrees of tourism development: the coast, with established forms of seaside tourism and mass-market; the countryside with high potential for the development of alternative and niche forms of tourism (naturalistic, sports, food and wine), but with spontaneous and unstructured forms of tourism',

which emphasizes the dual vocation of the area, the seaside and the rural one, unfortunately little valued due to a lack of shared management at the local level. This means that the destination is unable to emerge on the market by fully expressing its resources because it would need a meta-governance organization that can define its identity and communicate it to the market.

Precisely on this latter point, the actors highlighted that in addition to the possible experiences related to the sea and the coast, the countryside has a high potential for the development of alternative and niche forms of tourism (e.g. nature, sports, food and wine) that currently represent tourist experiences that are spontaneously proposed by individual actors and that are not integrated with other structures in the area.

The participants at the tables showed much more sharing and alignment in communicating and transferring the strengths of the territory; in particular the image of the destination, the heritage of natural resources and the presence of small associations that offer experiential tourist services (e.g. snorkelling, trekking, cooking class, events) but which at the same time are not co-ordinated with each other. There is a risk of duplication and overlap by confirming that the unexpressed resources and the human heritage, while representing an opportunity to expand market segments, remain an issue that must be addressed at the governance level, and remain the main critical issues for relaunching the destination.

4.5 Discussion and Practical Implications for Future Development

The main purpose of this research is to define a repositioning model for demand-driven destinations starting from a participatory planning process involving local stakeholders. Despite the exploratory nature of the analysis, interesting insights emerge. According to Quinlan *et al.* (2013), this research is a first attempt to engage local stakeholders in a shared governance project through a participatory planning process. The high level of interest in participation and what emerged from the meetings seems to show good awareness and the first level of engagement achieved, by confirming the theoretical framework of stakeholder engagement. In this explorative stage, it emerges from the tables that there is a latent need for destination governance that favours sharing between the actors, enhances the destination in all its peculiarities and sets strategic objectives to intercept the new demand paradigms (Moscardo, 2011; Hoving, 2017; Sigala, 2019; Stankov *et al.*, 2020), thus favouring an extension of the season and avoiding summer overtourism.

From a strategic point of view, we have identified a need to set *strategic goals*. As the current tourism is highly generated by the summer tourist flows, resulting in overtourism and limited economic value for local stakeholders, a good option would be to enhance the peculiarities of the rural part of the destination to intercept new demand segments (digital detox tourism, mindfulness tourism), which is aligned to the latent need expressed by both operators and local administrators to lengthen the tourist season.

Even in a *network* pespective, and considering the findings that emerged before, what emerged is a latent and unspoken will to be guided by a central player that can give a strategic direction to the tourist development of the

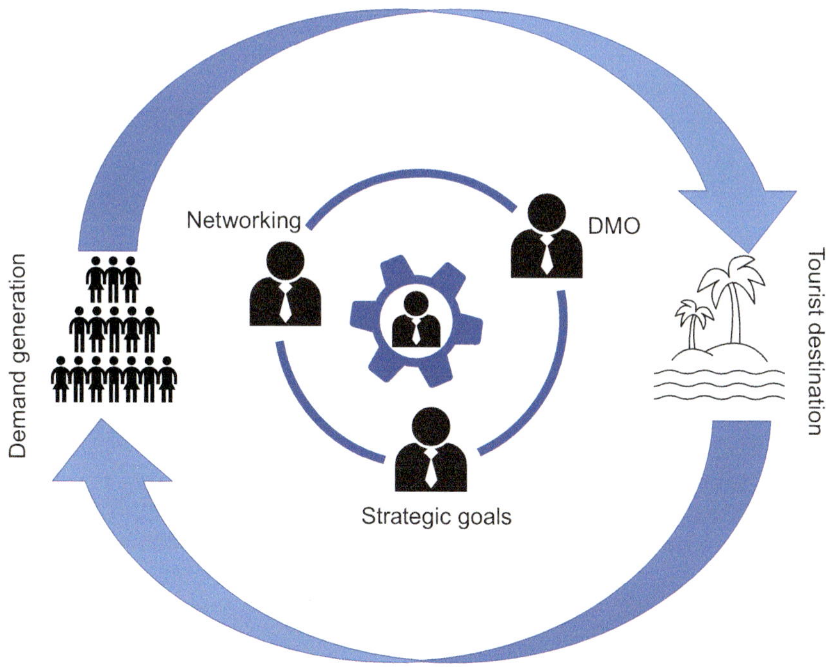

Fig. 4.1. Go-to-market framework.

destination. Although implied, we believe that from what was perceived during the meetings, the actors would like to be guided by a destination management organization. Indeed, the ability to enhance the multiple vocations of the territory and to intercept new demand seem to be associated with the capabilities that a meta-orchestrator (a DMO) can have in reconciling market needs with those of the destination.

Finally, a positive attitude in favour of guided sharing was found. It can be said that participatory planning with thematic tables has given a stimulus and a role to each player in the area. Therefore, it might be important to continue this engagement process with the phases of *integration* to reinforce the interactive networking and develop positive relationships with the *management* stage by establishing activities of monitoring that enhance practice and improve collaborative results (Waligo *et al.*, 2013).

Our study has practical and academic implications in the field of tourism and destination marketing. Indeed, the analysis of the tourist scenario and the results obtained after the thematic tables identify some possible go-to-market strategies. Firstly, the decongestion of the coast and the lengthening of the season, which emerged as priorities from the analysis carried out, could be accomplished by designing new tourism experiences linked to the resources of the countryside, which would complement and/or be in addition to the 'beach holidays' product, and offer such experiences to current tourists. Coherently to the recent demand trends (Moscardo, 2011; Hoving, 2017; Sigala, 2019; Stankov *et al.*, 2020), in this regard local actors could focus on:

Sports activities. These are experiences related to the outdoors (hiking, mountain-biking, paragliding, climbing). The design effort should be directed towards involving sports associations and establishing and promoting the possibilities already offered in the area.

Slow tourism. These tourism experiences are related to the 'slow' use of the area through cycling or hiking. The design effort should be oriented towards enhancing existing itineraries or creating new itineraries (based on cycle paths or paths for walking) and should involve hospitality and food-service operators oriented towards this form of tourism.

Cultural short breaks. These are experiences that enhance both the offer of artistic places in the countryside area and the so-called 'cultural deposits' of the Cilento area. The design effort should be directed towards systematizing the existing offers and should involve hospitality and transportation service operators to organize short tours of the area.

Detox. These tourism experiences, which are showing strong growth, are linked to moments of detachment from the frenzy of modern life and the spasmodic use of digital devices. The design effort should be directed towards involving hospitality operators that are able to offer moments of relaxation in out-of-the-ordinary contexts (for example, monasteries and farmhouses).

Finally, regarding the creation of new tourism products for new market segments, the ability to expand the territorial offer in terms of both the markets served and tourism experiences could reach maturity through the development of a portfolio of different tourist experiences linked both to the sea and to the resources of the hinterland, independent of each other and economically self-sufficient. This portfolio could be offered to the Italian and international market through multiple distribution channels, allowing the destination area to attract high-quality and high-spending tourism over a period that goes beyond the mere summer season.

4.6 Concluding Remarks

This work demonstrates how stakeholder engagement in a participatory planning process can outline the strengths and weaknesses of a tourist destination. In particular, the case analysed shows that a bottom-up perspective may facilitate dialogue and lead people to converge on shared priority objectives for intervention in a destination without formalized governance. Moreover, a form of engagement such as that presented in this chapter is an opportunity for discussion with operators working under a market logic and an opportunity to illustrate emerging trends to them. Finally, this research shows that the stage of attraction is the most important for capturing the interest of local stakeholders and their loyalty to build a cohesive group of people aligned to achieve shared goals.

Especially in a demand-driven destination such as Centola-Palinuro, the experience of thematic meetings may shed light on the problem setting and help find solutions and opportunities together to enhance the potential of the destination. Thematic meetings are also an opportunity to create a network and to co-create value for the destination.

From a theoretical perspective, this research enriches the existing literature on stakeholder engagement with an interesting case study in the tourism field, which is an ideal setting to involve stakeholders around a shared goal. Secondly, in the field of tourism motivation, it helps show how demand trends could represent strategic goals for a tourism destination. Finally, it confirms the theoretical framework of Quinlan *et al.* (2013) by showing the positive effect of attracting stakeholders through a participatory process.

Especially after a pandemic crisis that has had a heavy negative impact on tourism, this study constitutes a tangible document from which to start reformulating a destination's offer. In particular, the highlighted demand trends represent some interesting paradigms emphasized by the Covid-19 crisis, such as mindfulness, digital detox and nature-based active vacations. We believe that destinations could start to design new experiences based on benefits and spiritual values such as those represented by these paradigms. Indeed, in this last year, people are experiencing stay-at-home, social distancing and the digitalization of most of the daily activities (school, work, meetings), so that the need to have a break to rest is perhaps more felt than before. Nature, escape and mindfulness could be the paradigms for the new normality (Sibi, 2020), so there is a need for the destinations to reinvent their value propositions by designing their offering and communication on experience and quality (Bazzanella *et al.*, 2020), guaranteeing safety and finding a new way to generate value and distribute income in the community.

Some limitations need to be considered in future research. First, a single case study was used; thus, the results cannot be generalized. Second, the thematic meetings took place before the pandemic crisis; therefore, new meetings should be organized to update the framework. Based on these limitations, it would be interesting to extend the study by repeating the thematic meetings with local stakeholders digitally to capture their actual feelings. The research could be improved by conducting a two-step analysis in which some operational activities identified in the matrix are carried out and the results monitored. In the next step, starting from the current results, data will be analysed with specific software (e.g. WordStat) to aggregate dimensions extracted and customized for our investigation.

References

Ansell, C. and Gash, A. (2008) Collaborative governance in theory and practice. *Journal of Public Administration Research and Theory* 18(4), 543–571. DOI: 10.1093/jopart/mum032.

Atorough, P. and Martin, A. (2012) The politics of destination marketing: assessing stakeholder interaction choice orientations toward a DMO formation, using the Thomas-Kilmann Conflict Mode Instrument. *Journal of Place Management and Development* 5, 35–55.

Batat, W. (2019) *The New Luxury Experience*. Springer International Publishing, Cham, Switzerland.

Bazzanella, A., Beritelli, P. and Grigolli, P. (eds) (2020) *Destination Management (R) Evolution: Il Trentino come laboratorio di innovazione turistica (2002–2020)*. FrancoAngeli, Milan, Italy.

Beritelli, P. (2011) Cooperation among prominent actors in a tourist destination. *Annals of Tourism Research* 38(2), 607–629. DOI: 10.1016/j.annals.2010.11.015.

Bigné, E., Gnoth, J. and Andreu, L. (2008) Advanced topics in tourism market segmentation. In: Woodside, A.G. and Martin, D. (eds) *Tourism Management: Analysis, Behaviour and Strategy*. CAB International, Wallingford, UK, pp. 151–173.

Bigné, J.E. and Andreu, L. (2004) Emotions in segmentation: an empirical study. *Annals of Tourism Research* 31, 682–696.

Bishop, S.R. *et al.* (2004) Mindfulness: a proposed operational definition. *Clinical Psychology: Science and Practice* 11(3), 230–241.

Bizirgianni, I. and Dionysopoulou, P. (2013) The influence of tourist trends of youth tourism through Social Media (SM) & Information and Communication Technologies. *Procedia – Social and Behavioral Sciences* 73(7–8), 652–660. DOI: 10.1016/j.sbspro.2013.02.102.

Bramwell, B. and Lane, B. (2011) Critical research on the governance of tourism and sustainability. *Journal of Sustainable Tourism* 19(4-5), 411–421. DOI: 10.1080/09669582.2011.580586.

Bramwell, B. and Sharman, A. (1999) Collaboration in local tourism policymaking. *Annals of Tourism Research* 26(2), 392–415. DOI: 10.1016/S0160-7383(98)00105-4.

Buffa, F. (2015) Young tourists and sustainability: profiles, attitudes, and implications for destination strategies. *Sustainability* 7(10), 14042–14062. DOI: 10.3390/su71014042.

Byrd, E.T. (2007) Stakeholders in sustainable tourism development and their roles: applying stakeholder theory to sustainable tourism development. *Tourism Review* 62(2), 6–13. DOI: 10.1108/16605370780000309.

Cabiddu, F., Lui, T.-W. and Piccoli, G. (2013) Managing value co-creation in the tourism industry. *Annals of Tourism Research* 42(1), 86–107. DOI: 10.1016/j.annals.2013.01.001.

Calza, F., Ferretti, M., Risitano, M. and Sorrentino, A. (2019a) Analysing customer experience in heritage tourism: empirical evidence from an Italian cultural district. *Mercati & competitivit* 3.

Calza, F., Simoni, M., Sorrentino, A., Testa, G. and Trunfio, M. (2019b) I processi di pianificazione partecipata nelle destinazioni demand-driven. Il caso del piano strategico del turismo di Centola-Palinuro. XXIII Edizione del Rapporto sul Turismo Italiano, Rogiosi Editore.

Carciofi, A. (2017) *Digital Detox: Focus & produttivit per il manager nell'era delle distrazioni digitali.* Hoepli editore.

Chen, L.I.-L., Scott, N. and Benckendorff, P. (2017) Mindful tourist experiences: a Buddhist perspective. *Annals of Tourism Research* 64(2), 1–12. DOI: 10.1016/j.annals.2017.01.013.

Choibamroong, T., Laws, E., Richins, H., Agrusa, J. and Scott, N. (2011) A stakeholder approach for sustainable community-based rural tourism development in Thailand. In: Laws, E., Richins, H., Argusa, J. and Scott, N. (eds) *Tourist Destination Governance: Practice, Theory and Issues*. CAB International, Wallingford, UK, pp. 173–186.

da Mota, V.T. and Pickering, C. (2020) Using social media to assess nature-based tourism: current research and future trends. *Journal of Outdoor Recreation and Tourism* 30(2), 100295. DOI: 10.1016/j.jort.2020.100295.

Dogru, T., Sirakaya-Turk, E. and Crouch, G.I. (2017) Remodeling international tourism demand: old theory and new evidence. *Tourism Management* 60(5), 47–55. DOI: 10.1016/j.tourman.2016.11.010.

Egger, I., Lei, S.I. and Wassler, P. (2020) Digital free tourism: an exploratory study of tourist motivations. *Tourism Management* 79(5), 104098. DOI: 10.1016/j.tourman.2020.104098.

Eisenhardt, K.M. (1989) Building theories from case study research. *Academy of Management Review* 14(4), 532–550. DOI: 10.5465/amr.1989.4308385.

Ejarque, J. (2003) *La Destinazione Turistica di Successo: Marketing e Management*. Hoepli, Milan, Italy.

Ejarque, J. (2009) *Destination Marketing: La Nuova Frontiera Della Promocommercializzazione Turistica*. Hoepli, Milan, Italy.

Ferraresi, M. and Schmitt, B.H. (2018) *Marketing Esperienziale: Come Sviluppare L'esperienza Di Consumo*. FrancoAngeli, Milan, Italy.

Frauman, E. and Norman, W.C. (2004) Mindfulness as a tool for managing visitors to tourism destinations. *Journal of Travel Research* 42(4), 381–389. DOI: 10.1177/0047287504263033.

Go, F.M. and Trunfio, M. (2012) A paradigm shift from tourism destination management to democratic governance of place branding. *Journal of Travel & Tourism Research* 12, 4–17.

Go, F.M. Trunfio, M. and Lucia, M.D. (2013) Social capital and governance for sustainable rural development. *Studies in Agricultural Economics* 115(2), 104–110. DOI: 10.7896/j.1220.

Holbrook, M.B. and Hirschman, E.C. (1982) The experiential aspects of consumption: Consumer fantasies, feelings, and fun. *Journal of Consumer Research* 9(2), 132–140.

Hoving, K. (2017) *Digital Detox Tourism: Why Disconnect? What Are the Motives of Dutch Tourists to Undertake a Digital Detox Holiday?* Umea University, Sweden.

Kirwin, M., Harper, N.J., Young, T. and Itzvan, I. (2019) Mindful adventures: a pilot study of the outward bound mindfulness program. *Journal of Outdoor and Environmental Education* 22(1), 75–90. DOI: 10.1007/s42322-019-00031-9.

Lew, A.A. (2008) Long tail tourism: new geographies for marketing niche tourism products. *Journal of Travel & Tourism Marketing* 25(3-4), 409–419. DOI: 10.1080/10548400802508515.

Loureiro, S.M.C., Stylos, N. and Miranda, F.J. (2019) Exploring how mindfulness may enhance perceived value of travel experience. *The Service Industries Journal* 40(11-12), 800–824. DOI: 10.1080/02642069.2019.1600672.

Lucia, M.D. and Buffa, F. (2011) Il ruolo dell'attore pubblico e della destination management organisation nel marketing delle destinazioni community-type. Uno studio esplorativo. Mercati e Competitivit.

Mahika, E.C. (2011) Current trends in tourist motivation. *Cactus Tourism Journal* 2, 15–24.

March, R. and Wilkinson, I. (2009) Conceptual tools for evaluating tourism partnerships. *Tourism Management* 30(3), 455–462. DOI: 10.1016/j.tourman.2008.09.001.

Martini, U. (ed.) (2017) *Management e Marketing Delle Destinazioni Turistiche Territoriali: Metodi, Approcci e Strumenti*. McGraw-Hill Education.

Mehmetoglu, M. and Engen, M. (2011) Pine and Gilmore's concept of experience economy and its dimensions: An empirical examination in tourism. *Journal of Quality Assurance in Hospitality & Tourism* 12(4), 237–255.

Miao, L., Im, J., Fu, X., Kim, H. and Zhang, Y.E. (2021) Proximal and distal post-COVID travel behavior. *Annals of Tourism Research* 88(9), 103159. DOI: 10.1016/j.annals.2021.103159.

Miles, M.B. and Huberman, A.M. (1994) *Qualitative Data Analysis: An Expanded Sourcebook*. Sage, Thousand Oaks, California.

Millán, A., Fanjul, M.L. and Moital, M. (2016) Segmenting the business traveler based on emotions, satisfaction and behavioral intention. *Psychology & Marketing* 33(2), 82–93. DOI: 10.1002/mar.20856.

Moscardo, G. (2009) Understanding tourist experience through mindfulness theory. In: Kozak, M. and Decrop, A. (eds) *Handbook of Tourist Behavior: Theory & Practice*. Routledge, New York, pp. 99–115.

Moscardo, G. (2011) Searching for well-being: exploring change in tourist motivation. *Tourism Recreation Research* 36(1), 15–26. DOI: 10.1080/02508281.2011.11081656.

Pechlaner, H., Raich, F., Beritelli, P., Franch, M., Martini, U. *et al.* (2010) Roles and opinions of primary and secondary stakeholders within community-type destinations. *Tourism Review* 65, 74–86.

Pesonen, J. and Komppula, R. (2010) Rural wellbeing tourism: motivations and expectations. *Journal of Hospitality and Tourism Management* 17(1), 150–157. DOI: 10.1375/jhtm.17.1.150.

Plaza-Ubeda, J.A., de Burgos-Jimenez, J. and Carmona-Moreno, E. (2010) Measuring stakeholder integration: knowledge, interaction and adaptation behaviour dimensions. *Journal of Business Ethics* 3(3), 419–442. DOI: 10.1007/s10551-009-0231-9.

Presenza, A. and Formato, R. (2018) *Management Della Destinazione Turistica: Attori, Strategie e Indicatori Di Performance*. FrancoAngeli, Milan, Italy.

Pyke, J., Law, A., Jiang, M. and de Lacy, T. (2018) Learning from the locals: the role of stakeholder engagement in building tourism and community resilience. *Journal of Ecotourism* 17(3), 206–219. DOI: 10.1080/14724049.2018.1505586.

Quinlan, T., Lally, A.M. and O'Donovan, I. (2013) Stakeholder engagement in destination management: a systematic review of literature. In: Tourism and Hospitality Research in Ireland Conference-THRIC. Galway-Mayo Institute of Technology.

Richards, G. (2015) The new global nomads: youth travel in a globalizing world. *Tourism Recreation Research* 40(3), 340–352. DOI: 10.1080/02508281.2015.1075724.

Risitano, M., Romano, R. and Sorrentino, A. (2020) Segmenting cruise passengers from the experiential marketing perspective: an explorative study. *Research in Transportation Business & Management* 100590.

Rittichainuwat, B.N. (2018) *Special Interest Tourism*. Cambridge Scholars Publishing, Newcastle upon Tyne.

Ruhanen, L. (2009) Stakeholder participation in tourism destination planning another case of missing the point? *Tourism Recreation Research* 34(3), 283–294. DOI: 10.1080/02508281.2009.11081603.

Ruhanen, L., Scott, N., Ritchie, B. and Tkaczynski, A. (2010) Governance: a review and synthesis of the literature. *Tourism Review* 66(4), 4–16. DOI: 10.1108/16605371011093836.

Schmitt, B.H. (2010) *Customer Experience Management: A Revolutionary Approach to Connecting with Your Customers.* John Wiley & Sons, New Jersey.

Shen, S., Sotiriadis, M. and Zhang, Y. (2020) The influence of smart technologies on customer journey in tourist attractions within the smart tourism management framework. *Sustainability* 12(10), 4157. DOI: 10.3390/su12104157.

Sibi, P.S. (2020) Changing paradigms of travel motivations post COVID-19. *International Journal of Management* 11, 489–500.

Sigala, M. (2019) The bright and the dark sides of social media in tourism experiences, tourists' behavior, and well-being. In: Dallen, J.T. (ed.) *Handbook of Globalisation and Tourism.* Edward Elgar Publishing, Cheltenham, UK.

Smith, M.K. and Diekmann, A. (2017) Tourism and wellbeing. *Annals of Tourism Research* 66(5), 1–13. DOI: 10.1016/j.annals.2017.05.006.

Stankov, U., Filimonau, V. and Vujičić, M.D. (2020) A mindful shift: an opportunity for mindfulness-driven tourism in a post-pandemic world. *Tourism Geographies* 22(3), 703–712. DOI: 10.1080/14616688.2020.1768432.

Stará, J. and Peterson, C. (2017) Understanding the concept of wellness for the future of the tourism industry: a literature review. *Journal of Tourism & Services* 8, 18–29.

Tribe, J., Dann, G. and Jamal, T. (2015) Paradigms in tourism research: a trialogue. *Tourism Recreation Research* 40(1), 28–47. DOI: 10.1080/02508281.2015.1008856.

Trunfio, M. and Della Lucia, M. (2019) Engaging destination stakeholders in the digital era: the best practice of Italian regional DMOs. *Journal of Hospitality & Tourism Research* 43(3), 349–373. DOI: 10.1177/1096348018807293.

Waligo, V.M., Clarke, J. and Hawkins, R. (2013) Implementing sustainable tourism: a multi-stakeholder involvement management framework. *Tourism Management* 36(1), 342–353. DOI: 10.1016/j.tourman.2012.10.008.

Yachin, J.M. (2018) The 'customer journey': learning from customers in tourism experience encounters. *Tourism Management Perspectives* 28(1), 201–210. DOI: 10.1016/j.tmp.2018.09.002.

Yin, R.K. (2017) *Case Study Research and Applications: Design and Methods.* Sage, Thousand Oaks, California.

Yousaf, A., Amin, I. and Santos, J.A.C. (2018) Tourists' motivations to travel: a theoretical perspective on the existing literature. *Tourism and Hospitality Management* 24(1), 197–211. DOI: 10.20867/thm.24.1.8.

Websites

http://www.eptsalerno.it/index.php/statistiche

https://www.beniculturali.it/mibac/multimedia/MiBAC/documents/1481892223634_PST_2017_IT.pdf

5 Comparison of International Tourist Profiles in the Spanish Wine and Olive Oil PDOs

Jesús Barreal Pernas and Gil Jannes*

Universidad Complutense de Madrid, Spain

5.1 Introduction

Spain was visited by around 83 million international tourists during 2019, with an economic impact of 92,278.26 million euros, which represents roughly 7.5% of Spanish GDP (INE, 2020). Therefore, international demand is crucial for both the tourism sector and the Spanish economy in general. In some regions, such as the Canary or Balearic Islands, tourism, together with its ancillary services, forms the main economic activity, thus causing a substantial socioeconomic dependence. In other areas, such as Madrid or the Basque Country, with a more diverse economic structure and traditionally less oriented towards international tourism, this still represents an important extra source of income.

Spain has many attractions for foreign visitors, both natural (climate and varied geography, from mountains to beaches) and cultural. Among Spain's cultural attractions, gastronomy is taking an increasingly important position, from the traditional sangría and tapas to the molecular gastronomy of the past decades (García-Segovia *et al.*, 2014). Spain's gastronomy is intimately linked to its role as a provider of high-quality agricultural and derived products, including wine and olive oil. This creates a potential for agrotourism which has been recognized long ago (see, e.g., Armesto López and Martin, 2006). Nevertheless, the agricultural and livestock sector is still the main economic activity in most agro-rural areas, with tourism gradually starting to play a more significant role, and thus providing an additional source of income and dynamizing the local economy.

Spain, like many other countries, has set special protection to some of its regional specialties through a system of Protected Designation of Origin (PDO). These high-quality goods present peculiar and notable

*Corresponding author: gjannes@ucm.es

DOI: 10.1079/9781789248753.0005

characteristics and are linked with a specific geographical area, similar to Modena balsamic vinegar, Champagne wine or Dorset Blue Vinney cheese in other countries. The PDOs offer an easily identifiable form of brand differentiation and can also have a significant impact on tourism demand (Armesto López and Martin, 2006). Spain currently has 90 wine PDOs and 102 other agro-alimentary PDOs. These involve a wide variety of food products including cheese, ham and other types of meat, honey, fruit, vegetables, olive oil and wine. These last two, olive oil and wine, form an essential economic sector in several Spanish regions. Indeed, Spain is a world-leading producer of wine and olive oil. According to OIV (2019), Spain is the third wine producer worldwide by volume (after Italy and France) and the first in terms of vineyard extension. However, the country only ranks eighth in the world in terms of consumption. The international wine markets absorb the surplus production, and indeed Spain is the number-one wine-exporting country by volume. On the other hand, Spain leads the ranking of olive oil-producing countries and has the largest land area focused on olive cultivation (IOC, 2020). As for wine, production largely surpasses the national consumption, and Spain is thus also the world's number-one olive oil-exporting country.

To sum up, Spain plays a world-leading role in both sectors, wine and olive oil, and these contribute importantly to Spain's international trade. This is not merely a consequence of the volume produced but also a recognition of their quality and characteristics. In spite of these leading roles in tourism, on the one hand, and in wine and olive oil production on the other, the interaction between them both still needs much improvement. Following research such as that of Hall and Mitchell (2007) or Slocum and Curtis (2017), who propose the tourist promotion of gastronomy and food products, the Spanish tourism sector could consider a stronger emphasis on wine and olive oil production as a tourist attraction.

With this objective in mind, the current research develops a segmentation among international tourists visiting PDO-protected regions of olive oil and wine production in Spain. In particular, this contribution aims to identify the different types of international tourists visiting such Spanish wine and olive oil PDOs, regardless of whether they actually visit a winery or an olive orchard and mill, in order to identify the potential for such types of tourism, rather than analysing the existing agro- and gastro-tourism. The profiles of international tourists visiting these regions (wine, olive oil, or mixed) are classified and compared in terms of country of origin, demographic and trip characteristics, motivation and activities carried out during the visit. The influence of tourists' expenditure on the probability of belonging to the different classes are also studied. This is carried out through a Latent Class Model (LCM). This method is well known in statistics, but has only relatively recently found its way into tourism segmentation research. LCM provides a very effective way of segmenting and comparing tourism profiles, and thus fits perfectly with the research aim just described.

The structure of this contribution is as follows. The next section contains a brief literature review. Section 3 deals with the methodology and data. Some general aspects of market segmentation are briefly outlined as is Latent Class Modelling, and the way in which the data were obtained and classified is described. The number, distribution and characteristics of the obtained classes, as well as the influence of the expenditures, are detailed in the results section (4). Section 5 contains some final comments and suggestions for further research.

5.2 Wine and Oil Tourism in Spain: A Literature Review

There is a large body of research on Spanish tourism, some of which focuses on gastronomy, agricultural or food production in tourism activities. Wine tourism in general, in the Mediterranean region and Spain in particular, is relatively well studied (see Sánchez *et al.*, 2017 for a bibliometric survey). A variety of methods have been applied to study the prospects and impact of wine tourism in socioeconomic, cultural, environmental or even ethnographic terms, for the whole Mediterranean region (Hall and Mitchell, 2000), Spain in general (López-Guzmán *et al.*, 2011; Gómez and Molina, 2012; Gómez *et al.*, 2015), or concrete regions of Spain (Clemente-Ricolfe *et al.*, 2012; Carrasco *et al.*, 2019).

Olive oil tourism or 'oleo-tourism', and research into it, are more recent. Apart from a few isolated articles, the thorough overview (Hernández-Mogollón *et al.*, 2019) cites 2010 as the birth year of research into oleo-tourism. Four years later, Millán *et al.* (2014) still called olive oil tourism a 'new market segment', while even very recently Pulido-Fernández *et al.* (2019) describe olive oil tourism as 'an emerging tourism typology'. Much of the research in this field focuses on the potential of olive oil tourism to contribute to rural development, and environmental and cultural preservation (see, e.g., Murgado, 2013; Millán *et al.*, 2014; Campón-Cerro *et al.*, 2014; López-Guzmán *et al.*, 2016; Agüera *et al.*, 2017; Arjona-Fuentes and Amador-Hidalgo, 2017). Olive oil tourism is seen as a factor to dynamize the rural economy while preserving the natural resources, as well as to protect the traditional heritage of olive oil production and consumption. Pulido-Fernández *et al.* (2019) have recently stressed the positive effects of this niche with respect to other tourism markets.

Most of the previous research, both in wine and olive oil tourism, considers eno-tourism and oleo-tourism directly and how these can be promoted. Few studies contemplate the PDOs as a touristic attractor to improve the local economy. A few exceptions are Kalenjuk *et al.* (2014) or Crescenzi and Mannelli (2015), who discuss the potential of PDOs to achieve better market differentiation and diversify the rural economy with tourism incomes. Finally, with the rare exception of Agudo (2010), no references have been found to any research contemplating a comparison between wine and olive oil tourism.

5.2.1 Market segmentation and Latent Class Modelling

Customer profile segmentation is crucial in any client-oriented branch of economics, and tourism is no exception (see, for example, Dolnicar, 2008 or the recent study of Goryushkina *et al.*, 2019). Market segmentation reveals public and private opportunities to attract specific customers by designing strategies according to target preferences. Kotler *et al.* (2017) identify three steps in customer-driven value-creation marketing strategies: the first step, segmentation, consists in identifying broad segments of potential customers; the second phase, targeting, selects the specific priority target groups based on their attractiveness (competitiveness, growth prospect and potential profitability) and the available resources; and in the last step, market positioning is defined and a specific marketing mix (price, product, distribution and promotion) is elaborated for each target group.

It should be noted that market segmentation is not a question that should necessarily be left only to businesses and companies. Segmentation can also be useful to help local or regional governments in determining political strategies (Barnett and Mahony, 2016). Thus, in rural areas, segmentation can contribute to defining public policies to attract international tourism and obtain an extra source of income for both tourist and agricultural sectors, while at the same time allowing a diversification of those local economies, which currently mainly depend on crop production. By contrast, in regions that have already a strong focus on international tourism, such as most of the Spanish Mediterranean coastline, market segmentation can assist public policies in attracting part of the tourism expenditure toward agrotourism, thereby achieving not only economic diversification but also improving the image and socioeconomic sustainability of tourism. Segmentation thus serves as a preliminary step for the design of targeting policies aimed at engaging the most attractive segments. Finally, after targeting, the lawmaker should develop a new legal framework to promote a sustainable marketing mix according to touristic policy goals.

In this contribution, market segmentation is implemented by using a Latent Class Model (LCM) with covariates. Latent structure models were developed in the 1950s (Lazarsfeld, 1950). A good modern summary with emphasis on its implementation in R through the poLCA package can be found in the study of Linzer and Lewis (2011). A more general reference for the R-project, which has been used to implement modelization, plots and analysis, is the R Core Team (R Core Team, 2019). Latent Class Modelization is a powerful statistical method which consists in analysing associations among the observed or 'manifest' variables, which are typically categorical, in order to segment the sample elements into a limited set of classes numbered by an unobserved or 'latent' variable. A typical example thus consists in analysing answers to a questionnaire about tourists' characteristics and interests (the manifest variables) in order to define market segments (the classes) of the respondents to the questionnaire (the sample elements). Latent Class Models have been used in tourism market segmentation studies. For example, Ramírez-Hurtado and Berbel-Pineda (2015) use an LCM to identify segments of tourists coming to Spain to play golf based on demographic and travel characteristics, motivation and satisfaction; Ferrer-Rosell *et al.* (2016) segment tourists according to the budget share they devote to different types of expenditure, while Li *et al.* (2020) perform an LCM to segment gambling tourists based on their perceived values and satisfaction with different gambling destinations.

In the context of wine tourism, many approaches to segmentation have been applied (see Alebaki and Iakovidou, 2011 for an overview and Molina *et al.*, 2015 for Spain). More recently, Cho *et al.* (2017) have developed a constraint-based approach to wine tourism segmentation, while Gu *et al.* (2018) use cluster segmentation to classify Chinese wine tourists in Australian wineries. However, the authors are aware of only two publications that apply an LCM to wine tourism, namely Bruwer and Li (2017) and Sánchez-Rivero and Pulido-Fernández (2012). Both of these, in fact, focus on methodological aspects of latent class tools in the context of tourism research, with wine tourism mostly used as an example rather than a research goal by itself. With respect to oleo-tourism, the authors can be very brief: there is no awareness of any segmentation study about olive oil tourism, let alone one based on a Latent Class Model. Since the LCM methodology is relatively novel in the context of agrotourism, the next section is dedicated to a brief summary of this method.

5.3 Method

The research baseline is the employment of statistical methods to perform a segmentation of all foreign visitors to Spanish mainland regions containing wine and/or olive oil PDOs. As discussed above, segmentation is a key element of core marketing strategy, as it allows the identification of relatively homogeneous demand segments within the overall heterogeneous group of candidate customers (in this case all foreign visitors to these regions) and defines positioning strategies for the target segments. The concrete segmentation method that is applied is a Latent Class Model, in which expenditures are included as covariates in order to study their effect on the probability of class membership.

Latent Class Modelling consists in a probabilistic categorization of a set of observations (for example answers of respondents) into classes, such that class membership produces expectations about the value of each observation for each of the manifest variables (i.e. how each respondent will answer each question). A key methodological component is that, within each class, the responses are assumed to be statistically independent. This allows the probability density function to be written as:

$$f\left(Y_i|\pi,p\right) = \sum_{r=1}^{R} p_r \prod_{j=1}^{J} \prod_{k=1}^{K_j} \left(\pi_{jrk}\right)^{Y_{ijk}} \quad (1)$$

where $j = 1,..., J$ indexes the manifest variables, and the function Y_{ijk} records each respondent's answer by defining $Y_{ijk} = 1$ if respondent i ($i = 1,..., N$) selects answer k ($k = 1,..., K_j$) for variable j and $Y_{ijk} = 0$ otherwise. P_r represents the proportion of respondents in category r (r = 1,..., R), and π_{jrk} is the conditional probability that a respondent in class r produces response k on variable j, such that $\sum_{k=1}^{K_j} \pi_{jrk} = 1$. The previous formula is easy to interpret by noting that, first, the construction $\prod (\pi_{jrk})\, Y_{ijk}$'selects' the values π_{jrk} for which $Y_{ijk} = 1$ (all other values being neutralized because of the exponent $Y_{ijk} = 0$). Furthermore, the assumption of conditional independence means that the joint probability over the different variables is simply the product of the individual probabilities. Finally, the summation gives a weighted sum over all classes.

The LCM is then determined by calculating estimators for p_r and π_{jrk}. In practice, this is performed by an iterative maximum-likelihood procedure. The *a posteriori* probability that an individual i with manifest variable values Y_i belongs to class r can then be found through Bayes' formula to be:

$$\hat{P}\left(r_i|Y_i\right) = \frac{\hat{p}_r f(Y_i;\hat{\pi}_r)}{\sum_{q=1}^{R} \hat{p}_q f(Y_i;\hat{\pi}_q)} \tag{2}$$

The previous model can be extended into a Latent Class Regression Model (LCRM) by the inclusion of a number of external factors or covariates X_i (in this research, the expenditure) which serve as predictors for the class membership probability of each sample element. To this effect, the class proportions p_r are replaced by probabilities p_{ri} for each element i to belong to class r, depending on the covariate values for that element. The extension from the case without covariates is straightforwardly obtained from (2) by replacing p by $p = (X_i; \hat{\beta})$, where b is the covariate coefficient vector (see Linzer and Lewis, 2011 for details).

Note that the LCM does not itself determine the number of classes. To select this number, a balance must be found between goodness of fit (higher number of classes) on the one hand and computational simplicity (lower number of classes) and the avoidance of overfitting on the other, while always keeping explanatory power as a main objective (Nylund *et al.*, 2007; Masyn, 2013). Two widely used criteria that penalize overfitting are the Akaike Information Criterion or AIC (Akaike, 1973) and the Bayesian Information Criterion or BIC (Schwarz, 1978):

$$\begin{aligned} AIC &= -2\Lambda + 2\phi \\ BIC &= -2\Lambda + \phi \ln N \end{aligned} \tag{3}$$

where Λ is the maximum log-likelihood and $\Phi = R_j\left(K_j - 1\right) + \left(R - 1\right)$ is the number of parameters to be estimated by the model. These criteria calculate the amount of information that is being lost between the full set of data and the model, and are therefore both minimization criteria. However, in practice one should not necessarily look for the absolute minimum but for the number that marks the onset where the decrease in AIC or BIC becomes non-substantial, i.e. when flattening starts. Also, it is important to verify the consistency of the results, namely whether the obtained groups are clearly differentiated and internally relatively homogeneous.

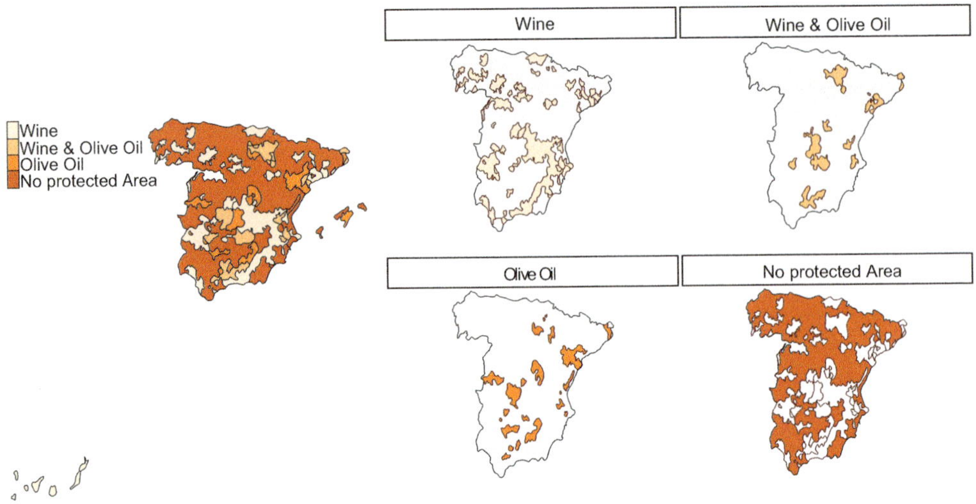

Fig. 5.1. Study area and types of regions. (Author's elaboration based on data from MAPA, 2020)

5.3.1 Data

Data were obtained from the Spanish National Statistics Institute, which records information from international tourists through direct questionnaires (INE, 2020). These questionnaires include questions about the journey, motivations and expenditure, as well as some personal information. The present research uses micro-data from September 2015 to July 2019, since the data from before September 2015, although available through the National Tourism Office, use a different methodology and question set. The Statistics Institute organizes the survey in all major airports, border-crossing points and tourist areas throughout the whole year. In this study, only those questionnaires have been analysed in which the tourists declare that they have spent a significant part of their stay in a wine and/or olive oil PDO. Hence, the paper uses information about the geographic distribution of Spanish PDOs to obtain the councils that belong to each protected area (MAPA, 2020). Concretely, only destinations within the following three categories are retained: wine PDOs, olive oil PDOs and mixed zones with both wine and olive oil PDOs. Fig. 5.1 plots these areas.

Note that this purely geographic method implies that many of the respondents that are taken into account are perhaps not specifically interested in wine or olive oil PDOs. Moreover, the Balearic and Canary islands are not included in the study, even though these have several wine and olive oil PDOs, for the following reason: there is a large number of responses from PDO regions on these islands, larger than all the other Spanish PDO regions taken together. However, the island tourist is still mainly of the 'sea and sun' type and shows a relatively low interest in gastronomic and cultural activities, thus skewing the overall results that could be obtained from our statistical analysis. Obviously, this does not imply that the question of PDO tourism in these islands is of no interest; on the contrary, convincing even only a small percentage of the large number of island tourists to spend just a single day on wine or olive oil tourism could have a considerable positive impact. But this question will be reserved for future study and the present study focuses on the mainland PDOs.

A total of 153,516 questionnaires were analysed from wine and/or olive oil protected areas during the study period. From these, 127,691 (83.2%) are from wine (only) PDOs; 12,548 (8.2%) from olive oil (only) PDOs; the remaining 13,277 (8.6%) from regions with both a wine and an olive oil PDO. The large preponderance of wine PDO responses can be explained by the larger involved area and the presence of traditional tourist zones, such as the Andalusian coastline around Cádiz, Marbella and Málaga, or the region of Alicante, which holds several wine PDOs.

In order to perform the actual market segmentation, Kotler *et al.* (2017) suggest the use of geographic, demographic, psychographic and behavioural variables to perform this market segmentation. Adapting this to the available data, the descriptors that are used here are: geographic; demographic; trip characteristics; and motivation and activities. The first item records the country of origin of the respondent. The second item considers age and occupational status. In the trip characteristics, the season, travelling party, accommodation type and length of stay are analysed. Finally, the last item records the primary motivation to visit Spain and the type of activities carried out (gastronomic, cultural, rural and leisure activities) during their visit. All these characteristics are used as manifest variables that form the basis of the LCM clustering, and they are summarized in Table 5.1. Note that the pool of potential target customers in this research is formed by all international visitors to a region containing a PDO, independently of whether they are *a priori* interested in participating in agrotourism or not. The idea is thus precisely to determine which classes among these tourists could be interesting targets for wine and olive oil tourism promotion.

The segmentation process itself, implemented through the basic LCM described above, helps to identify relatively homogeneous demand groups. Once these groups have been established, the tourist expenditures are analysed and related to the probability of pertaining to one class or another through an LCRM regression model. In particular, expenses in the following categories are considered: activities (including shows); goods and products for daily consumption; shopping for clothes, jewellery, crafts and other souvenirs; and mobility. These could be useful to define adequate product packages and marketing mixes, and thus optimize market positions. The expenditure analysis is thus complementary to the market segmentation itself and becomes important when transforming segmentation into targeting and positioning. The whole process is summarized in Fig. 5.2.

This expenditure is taken into account as daily averages, in other words, the questionnaire asks the tourists for their total expenses, which are then

Table 5.1. Manifest variables: summary of data. All numbers are percentages with respect to the total amount of responses.

Class	Variables	Wine PDOs	Mixed PDOs	Oil PDOs
Geographical origin	UK	18.94	14.72	15.05
	Germany	11.79	9.22	9.48
	France	11.13	29.70	29.25
	Italy	7.00	4.79	5.26
	The Netherlands	5.09	7.30	6.85
	Europe (other) & MidAsia	28.33	27.20	28.14
	Other	17.72	7.07	5.98
Demographical age	<24	10.96	8.50	8.65
	25–44	43.36	38.99	40.99
	45–64	34.26	37.23	35.77
	>65	11.43	15.28	14.58
Occupational situation	Employed	75.23	71.27	71.63
	Retired	13.79	18.60	17.90
	Student & other	10.99	10.13	10.47
Trip characteristics season	Winter	24.38	15.03	16.92
	Autumn	26.37	19.64	20.39
	Spring	24.76	29.18	28.68
	Summer	24.49	36.16	34.01
Travelling party	Alone	37.60	30.01	32.67
	Couple	34.77	34.74	32.77
	Family or Friends	27.63	35.25	34.56
Accommodation type	Hotel	55.81	37.72	36.09
	Other Market Accomod.	10.18	20.07	17.88
	No Market Accomod.	34.01	42.21	46.02
Length of stay	<3 days	19.00	13.59	14.18
	3–7 days	44.73	38.30	38.02
	1–2 Weeks	24.20	31.18	30.63
	>2 Weeks	12.07	16.93	17.17
Motivation & activities motivation	Sun & beach	34.81	35.12	33.58
	Leisure	24.61	24.03	21.28
	Family	13.78	16.95	21.78
	Culture	13.27	12.15	10.23
	Business	9.47	5.04	6.45
	Other	4.07	6.72	6.68

Continued

Table 5.1. Continued

Class	Variables	Wine PDOs	Mixed PDOs	Oil PDOs
Gastronomic	No	83.57	76.55	76.31
	Yes	16.43	23.45	23.69
Cultural	No	34.35	28.55	32.79
	Yes	65.65	71.45	67.21
Rural	No	76.70	67.94	69.33
	Yes	23.30	32.06	30.67
Leisure	No	60.92	61.36	60.08
	Yes	39.08	38.64	39.92

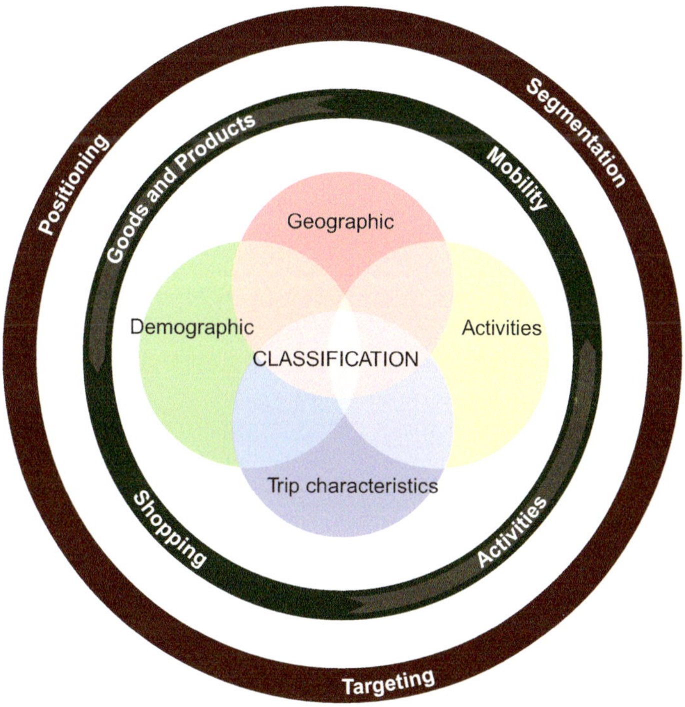

Fig. 5.2. Segmentation framework.

divided by the length of stay. Note that these expenses are included as co-variates. They do not therefore determine the LCM classes but are used afterwards as predictors for class membership. Note also that the respondents could leave one or more expenditure categories open, which explains why the box plots in Fig. 5.3 all start their range at zero. Finally, all prices have been calculated at 2016 price-equivalence by using the Consumer Price Index.

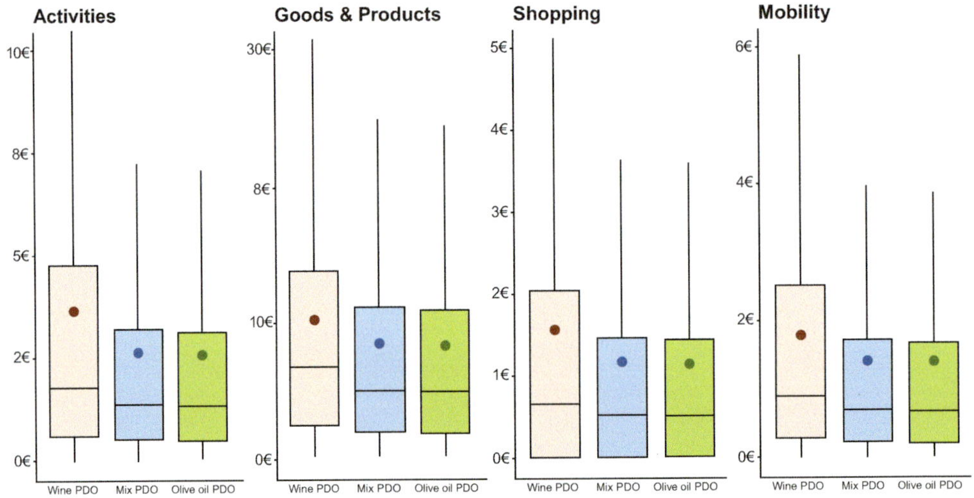

Fig. 5.3. Covariates: average daily expenses (in €) per person on activities, goods & products, shopping and mobility, separated per type of region. The box plots represent quartiles, with the dot depicting the mean in that particular category.

5.4 Results

5.4.1 Number, distribution and characteristics of classes

As described above, the number of classes is obtained from the AIC and BIC criteria, the latter of which is plotted in Fig. 5.4 (the AIC gives similar results). There is no absolute minimum in the range of observation, but downward saturation sets in for most scenarios at $n = 4$ classes, which are taken as a tentative conclusion, to be confirmed by the consistency of the obtained classes. Note that the result is perhaps less clear-cut for the wine PDOs but, in any

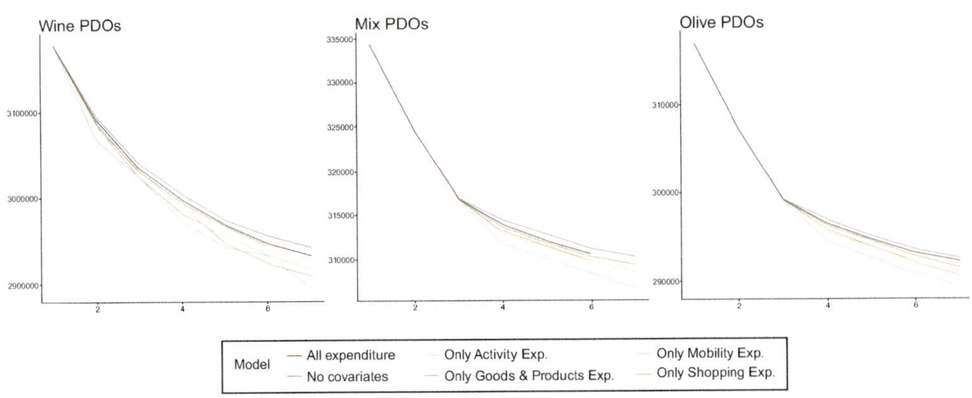

Fig. 5.4. Selection model: Bayesian Information Criterion.

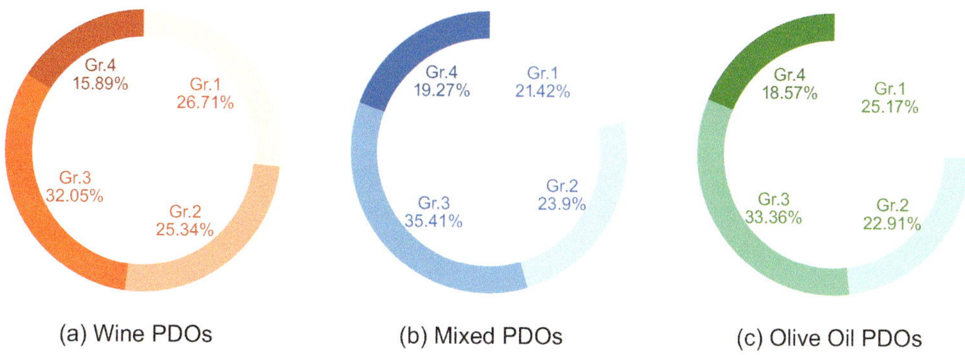

Fig. 5.5. Class percentage per type of region.

case, it is useful to take $n = 4$ there also since this will allow comparison of the obtained classes with the other types of regions.

The LCM analysis is thus performed with $n = 4$ classes, and the distribution of respondents over the obtained classes is depicted in Fig. 5.5. The two most salient features are the following. First, the respondents are comparably distributed over the four classes, which is a first confirmation that the choice $n = 4$ is adequate. Second, the percentages are quite similar for the three types of region, with the remark that there is a slightly higher percentage of Class 1 visitors in wine regions and slightly lower of Class 4. As seen below, the classes coincide quite well between the different types of region, thus allowing a direct comparison.

The composition of the classes is shown in Fig. 5.6. These can be characterized as follows:

> Class 1 consists mainly of students and young workers. They typically travel alone or with friends, preferably during the summer. A large proportion of them travels for leisure and visits family, which probably explains that they tend to use non-market accommodation. They stay relatively long (almost 70% for at least one week) and have a slightly higher than average participation in gastronomy.
>
> Class 2 consists of young and middle-aged employees who travel alone, their motives including (but not limited to) business. Their stay is typically rather short (less than a week) and they have a lower-than-average interest in (or time for) culture, rural activities and leisure.
>
> Class 3 has a similar demographic profile to Class 2, but within this class they are on holiday as couples, or with family or friends. Their objectives are culture, sun and beach, and leisure in general. They tend to spend a relatively short stay (the large majority stay less than a week) in a hotel.
>
> Finally, Class 4 consists of retired persons and older employees who travel all year round, mainly in a couple, for a relatively long stay (70% stay more than a week). An important motivation for them is sun and beach, and they also show a slightly higher than average interest in rural activities, although (curiously) a slightly less than average interest in leisure.

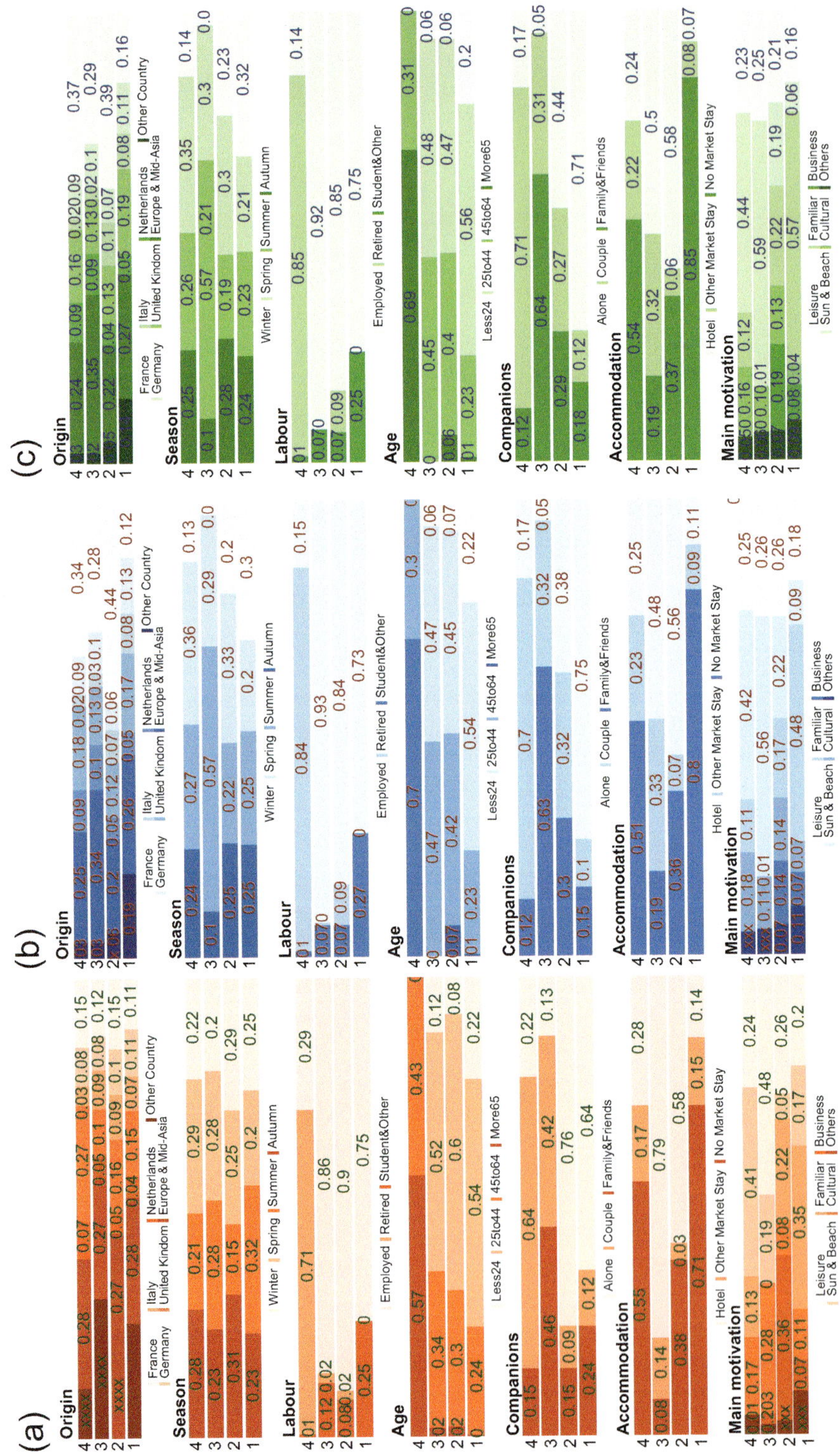
(a)
(b)
(c)
Origin
France | Germany | Italy | United Kindom | Netherlands | Europe & Mid-Asia | Other Country
Season
Winter | Spring | Summer | Autumn
Labour
Employed | Retired | Student&Other
Age
Less24 | 25to44 | 45to64 | More65
Companions
Alone | Couple | Family&Friends
Accommodation
Hotel | Other Market Stay | No Market Stay
Main motivation
Leisure | Sun & Beach | Familiar | Cultural | Business | Others

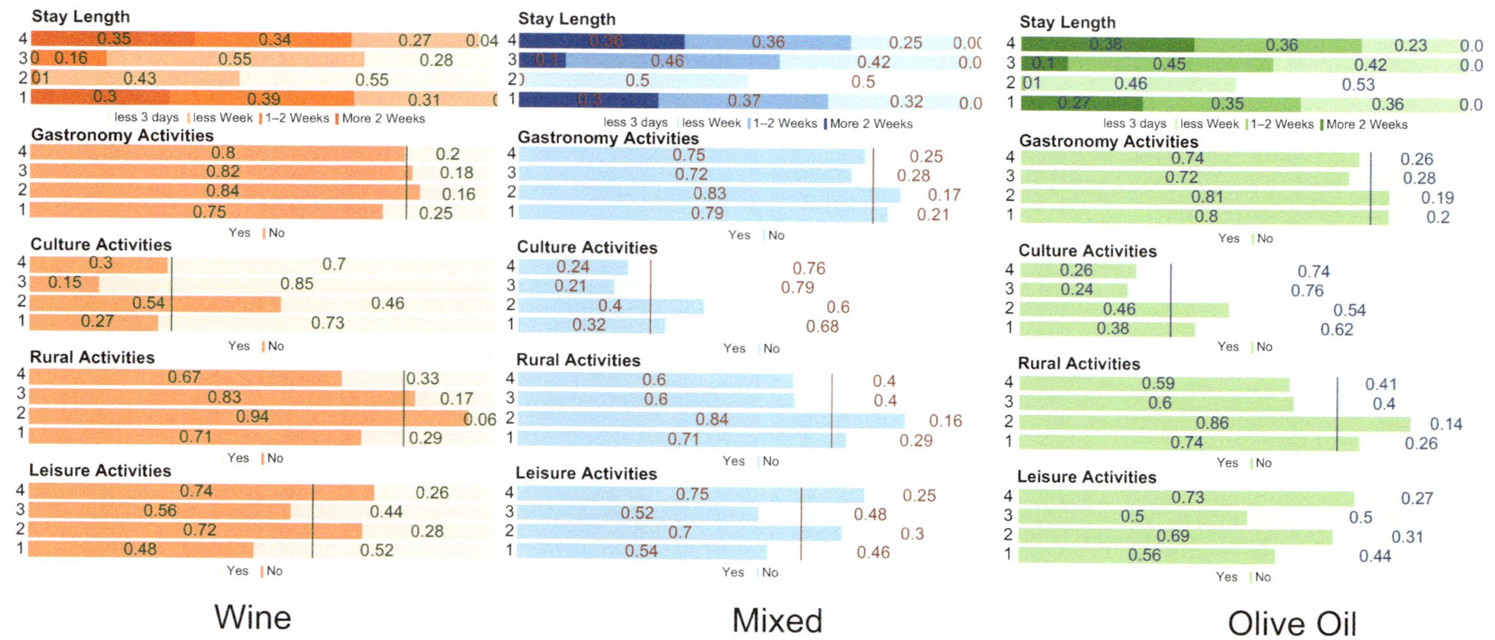

Fig 5.6. Class analysis, showing the proportion of answers for the variables of interest in each class and each type of region. The full vertical line in the lower four (dichotomic) variables indicates the overall average of yes/no answers in these categories.

Note that country of origin is not really a strong discriminator. Also, gender is left out because this is the gender of the respondent; most people do not travel alone, and among those who do, gender is not a determining factor for the overall profile. Finally, note that these classes are quite homogeneous across the different types of PDO. In fact, they are extremely similar between mixed (centre) and olive oil only (right) PDOs, although there are some interesting differences between wine PDOs and mixed/olive oil PDOs within each class. In particular, we note the following differences: for Classes 2, 3 and 4, there is a substantially larger percentage of French visitors to olive oil and mixed than to wine PDOs. For these same three classes (2, 3 and 4), there are slightly more autumn visitors to wine regions. Summer visitors belonging to Class 3 visit mixed and olive oil regions much more than wine regions. When Class 2 visitors go to wine regions, they tend to go alone and for business reasons in a much higher proportion than when they go to mixed/olive oil regions. Class 2 and especially 3 have a 'sun and beach' motivation much more often for olive and mixed regions than for wine regions. This seems to contradict the earlier remark that wine PDOs are more strongly represented in tourist coastal regions, and there is not currently a good interpretation for this curious fact.

Finally, in terms of interests and activities, the most important point to stress is the average levels of interest in the different types of activities (see Table 5.1; the average values are also marked in Fig. 5.6). A strong majority of the visitors to all types of region is interested in culture and an important minority in leisure; but only a minority is interested or has participated in gastronomic or rural activities. Interest in these two categories (gastronomic and rural) is, in fact, even lower in wine regions than in olive oil and mixed regions. This effect is particularly outspoken for Class 3 visitors (whereas Class 1 shows the inverse tendency) and can probably be explained by the importance of tourist coastal areas among the wine regions, as pointed out above.

5.4.2 Influence of expenditures

A box plot of the average daily expenditures of the various classes is shown in Fig. 5.7. On average, Class 2 has the highest expenditures in all four expenditure categories. It has high variability, but nevertheless well over 50% of its data points lie above the overall mean in each single category of each region and, in fact, drag the overall means upwards (all other classes' means systematically lie below the overall mean). Class 2 should therefore be an obvious priority target for marketing. The problem is that Class 2 has also shown the lowest interest in gastronomic and rural activities (see Table 5.1). This could be simply due to the shortness of their stay (typically less a week) in combination with the fact that many of them visit Spain for business purposes. Convincing them to actually visit a wine or olive oil producer might therefore be complicated, but there might be other ways of taking advantage of their high expenditure pattern in the current context; for example, the promotion of agro-gastronomic restaurants in the city centres.

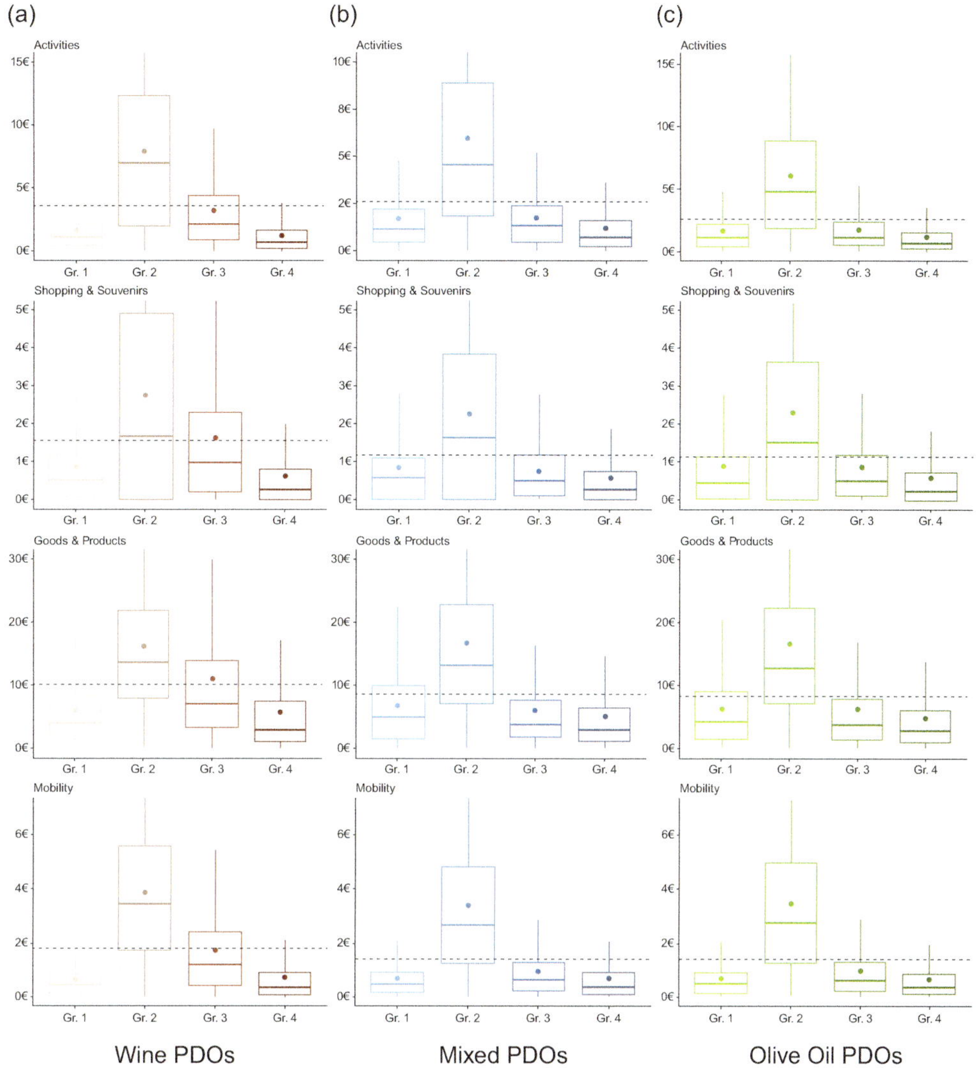

Fig. 5.7. Distribution of average daily expenditure per person in quartiles by class and expenditure category for the three types of regions. The dot indicates the mean expenditure in each class, while the dotted horizontal line indicates the overall average for each expenditure category.

Class 3 has a relatively interesting spending pattern in wine regions, but has also shown little interest in gastronomic and rural activities, especially when visiting wine regions. However, young and middle-aged employees who are on holiday as couples, or with family and friends, are traditionally among the largest group of winery visitors (Charters and Ali-Knight, 2002), and this is true also, in particular, in Spain (Molina *et al.*, 2015). It is therefore likely that local marketing efforts towards this group could be more fruitful, in the sense of convincing a larger percentage of visitors within this class to participate in eno-tourism.

Finally, Classes 1 and 4 spend little in general and show little variability, which is true also for Class 3 in olive oil and mixed regions. This does not mean that these classes should be left out of marketing efforts for PDO tourism, but they should perhaps not be targeted as specific priority groups, but rather integrated in the wider effort of promoting wine and olive oil PDO tourism.

The use of the expenditure in the 'activities' category as a predictor for class membership is shown in Fig. 5.8. This graph is based on the complete multi-variable LCM regression model, which is detailed for each type of region in Appendix A. Note that nearly all parameters of the model are highly significant in all regions, again confirming the choice of $n = 4$ classes. In order to isolate a single covariate, in this case the expenses in the 'activities' category, the multi-variable regression model requires fixing the remaining covariates. Three different scenarios are therefore contemplated: low, average and high spending in the remaining categories, corresponding to the overall first, second and third quartile values in each category. In all three scenarios, and for the three types of region, at low values of the activities expenditure, an intricate mixture of Classes 1, 3 and 4 is dominant. Class 4 starts decaying very rapidly, with Classes 1 and 3 holding on a bit longer (or even increasing in the low scenario). For higher activities expenses, Class 2 takes over and becomes completely dominant. Appendix B shows the same exercise, i.e. the evolution of class membership probability as a function of the other expenditure categories (goods and products, shopping and mobility). These graphs show patterns similar to Fig. 5.8.

5.5 Discussion and Conclusions

In this contribution, international tourists visiting Spanish wine or olive oil PDOs have been segmented through a Latent Class Model based on country

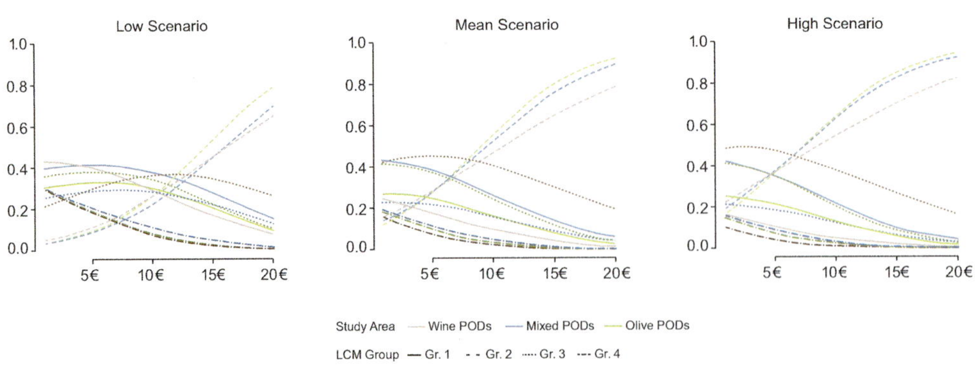

Fig. 5.8. Probability of class membership according to activities for three scenarios of other expenditure categories.

Hernández-Mogollón, J.M., Di-Clemente, E., Folgado-Fernández, J.A. and Campón-Cerro, A.M. (2019) Olive oil tourism: state of the art. *Tourism and Hospitality Management* 25(1), 179–207. DOI: 10.20867/thm.25.1.5.

INE (2020) Encuesta de gasto turístico. EGATUR. Instituto Nacional de Estadística. Available at: https://www.ine.es/dyngs/INEbase/es/operacion.htm?c=Estadistica_C&cid=1254736177002&menu=ultiDatos&idp=1254735576863 (accessed 26 November 2020).

IOC (2020) The world olive oil market in one click. International Olive Council [online resource]. Available at: https://www.internationaloliveoil.org/wp-content/uploads/2020/09/IOC-Olive-Oil-Dashboard-Septe html#production-2 (accessed 10 December 2020).

Kalenjuk, B., Tešanović, D., Gagić, S. and Knežević, M. (2014) Importance of food with protected origin and attitudes of employees towards the development of gastronomic tourism. *Tourism Education Studies and Practice* 3, 98–107. DOI: 10.13187/tesp.2014.3.98.

Kivela, J. and Crotts, J.C. (2006) Tourism and gastronomy: gastronomy's influence on how tourists experience a destination. *Journal of Hospitality & Tourism Research* 30(3), 354–377. DOI: 10.1177/1096348006286797.

Kotler, P., Bowen, J.T., Makens, J. and Baloglu, S. (2017) *Marketing for Hospitality and Tourism*. Pearson Education, Boston, Massachusetts.

Lazarsfeld, P.F. (1950) The logical and mathematical foundations of latent structure analysis. In: Stouffer, S.A. (ed.) *Measurement and Prediction*. John Wiley & Sons, New York, pp. 362–412.

Li, J., Bonn, M. and Kim, J.-H. (2020) A latent class segmentation analysis of gamblers in a gambling destination. *Journal of Destination Marketing & Management* 16(3), 100433. DOI: 10.1016/j.jdmm.2020.100433.

López-Guzmán, T., Rodríguez-García, J., Sánchez-Cañizares, S. and Luján-García, M.J. (2011) The development of wine tourism in Spain. *International Journal of Wine Business Research* 23(4), 374–386. DOI: 10.1108/17511061111186523.

López-Guzmán, T., Cañero Morales, P.M., Moral Cuadra, S. and Orgaz-Agüera, F. (2016) An exploratory study of olive tourism consumers. *Tourism and Hospitality Management* 22(1), 57–68. DOI: 10.20867/thm.22.1.1.

Linzer, D.A. and Lewis, J.B. (2011) poLCA: an R package for polytomous variable latent class anaysis. *Journal of Statistical Software* 42(10), 1–29. Available at: http://www.jstatsoft.org/v42/i10/ DOI: 10.18637/jss.v042.i10.

MAPA (2020) Zonas de Calidad Diferenciada: Denominaciones de Origen Protegidas (DOP) e Indicaciones Geográficas Protegidas (IGP). Ministerio de Agricultura, Pesca y Alimentación. Available at: https://www.miteco.gob.es/ide/metadatos/srv/spa/metadata.show?uuid=5210b5ac-557b-48d0-a8ef-138b08fbd970 (accessed 16 December 2020).

Masyn, K.E. (2013) Latent class analysis and finite mixture modeling. In: Little, T. (ed.) *Oxford Handbook of Quantitative Methods. Vol. 2: Statistical Analysis*. Oxford University Press, Oxford, UK, pp. 551–611.

Millán, G., Arjona, J.M. and Amador, L. (2014) A new market segment for olive oil: olive oil tourism in the south of Spain. *Agricultural Sciences* 5(03), 179–185. DOI: 10.4236/as.2014.53022.

Molina, A., Gómez, M., González-Díaz, B. and Esteban, Á. (2015) Market segmentation in wine tourism: strategies for wineries and destinations in Spain. *Journal of Wine Research* 26(3), 192–224. DOI: 10.1080/09571264.2015.1051218.

Murgado, E.M. (2013) Turning food into a gastronomic experience: olive oil tourism. *Options Méditerranéennes* 106, 97–109.

Nylund, K.L., Asparouhov, T. and Muthén, B.O. (2007) Deciding on the number of classes in latent class analysis and growth mixture modeling: a Monte Carlo simulation study. *Structural Equation Modeling: A Multidisciplinary Journal* 14(4), 535–569. DOI: 10.1080/10705510701575396.

OIV (2019) 2019 Statistical Report on World Vitiviniculture. International Organisation of Vine and Wine (OIV) Available at: http://www.oiv.int/public/medias/6782/oiv-2019-statistical-report-on- pdf (accessed 26 December 2020).

Ottenbacher, M.C. and Harrington, R.J. (2013) A case study of a culinary tourism campaign in Germany: implications for strategy making and successful implementation. *Journal of Hospitality & Tourism Research* 37, 3–28.

Pulido-Fernández, J.I., Casado-Montilla, J. and Carrillo-Hidalgo, I. (2019) Introducing olive-oil tourism as a special interest tourism. *Heliyon* 5(12), e02975. DOI: 10.1016/j.heliyon.2019.e02975.

R Core Team (2019) R: a language and environment for statistical computing. R Foundation for Statistical Computing, Vienna. Available at: https://www.R-project.org/ (accessed 26 January 2020).

Ramírez-Hurtado, J.M. and Berbel-Pineda, J.M. (2015) Identification of segments for overseas tourists playing golf in Spain: a latent class approach. *Journal of Hospitality Marketing & Management* 24(6), 652–680. DOI: 10.1080/19368623.2014.934980.

Sánchez-Rivero, M. and Pulido-Fernández, J.I. (2012) Testing heterogeneous image in cultural/non-cultural tourism markets: a Latent Model Approach. *International Journal of Tourism Research* 14(3), 250–268. DOI: 10.1002/jtr.850.

Sánchez, A.D., de la Cruz Del Río Rama, M. and García, J.Á. (2017) Bibliometric analysis of publications on wine tourism in the databases Scopus and WoS. *European Research on Management and Business Economics* 23(1), 8–15. DOI: 10.1016/j.iedeen.2016.02.001.

Santeramo, F., Seccia, A. and Nardone, G. (2017) The synergies of the Italian wine and tourism sectors. *Wine Economics and Policy* 6(1), 71–74. DOI: 10.1016/j.wep.2016.11.004.

Schwarz, G. (1978) Estimating the dimension of a model. *The Annuals of Statistics* 6(2), 461–464. DOI: 10.1214/aos/1176344136.

Skryl, T., Gregoric, M. and Dugi, V. (2018) Culinary trends in the Republic of Croatia as part of gastro tourism development. *European Research Studies Journal* 21(3), 465–475. DOI: 10.35808/ersj/1075.

Slocum, S.L. and Curtis, K.R. (2017) *Food and Agricultural Tourism: Theory and Best Practice.* Routledge, New York.

Thach, L. and Cogan-Marie, L. (2018) Wine tourism in Burgundy, France: an analysis of marketing practices. *Tourism Review International* 22(1), 81–95. DOI: 10.3727/154427218X15202734130468.

Williams, H.A. and Omar, M. (2014) Gastro-tourism as destination branding in emerging markets. *International Journal of Leisure and Tourism Marketing* 4(1), 1–18. DOI: 10.1504/IJLTM.2014.059257.

Appendix A: Latent Class Model Results

Wine PDOs, fit for four latent classes:

2/1	Coefficient $\beta r,i$	Std. error	t value	P (> |t|)
Intercept	–2.702	0.028	–97.9	0.000***
Activities	0.225	0.005	41.5	0.000***
Shopping	0.083	0.008	10.1	0.000***
Goods & products	0.053	0.001	35.8	0.000***
Mobility	0.757	0.012	63.8	0.000***
3/1	Coefficient $\beta r,i$	Std. error	t value	P (> |t|)
Intercept	–1.037	0.019	–54.3	0.000***
Activities	0.102	0.005	18.8	0.000***
Shopping	0.062	0.008	7.8	0.000***
Goods & products	0.041	0.001	30.0	0.000***
Mobility	0.529	0.012	45.3	0.000***
4/1	Coefficient $\beta r,i$	Std. error	t value	P (> |t|)
Intercept	–0.324	0.019	–17.2	0.000***
Activities	–0.104	0.007	–14.0	0.000***
Shopping	–0.213	0.011	–18.5	0.000***
Goods & products	–0.0002	0.001	–0.2	0.879
Mobility	0.174	0.014	12.7	0.000***
Maximum log-likelihood: –1485060				

Mixed PDOs, fit for four latent classes:

2/1	Coefficient $\beta r,i$	Std. error	t value	P (> |t|)
Intercept	–2.699	0.095	–28.5	0.000***
Activities	0.313	0.020	15.4	0.000***
Shopping	0.390	0.036	10.83	0.000***
Goods & products	0.090	0.005	17.6	0.000***
Mobility	0.597	0.033	18.2	0.000***
3/1	Coefficient $\beta r,i$	Std. error	t value	P (> |t|)
Intercept	–0.331	0.057	–5.8	0.000***
Activities	0.124	0.022	5.7	0.000***
Shopping	0.230	0.036	6.5	0.000***
Goods & products	0.030	0.005	5.9	0.000***
Mobility	–0.109	0.042	–2.5	0.011*
4/1	Coefficient $\beta r,i$	Std. error	t value	P (> |t|)
Intercept	0.099	0.048	2.1	0.037*
Activities	0.109	0.020	5.5	0.000***
Shopping	0.194	0.034	5.8	0.000***
Goods & products	0.012	0.004	2.7	0.000***
Mobility	0.186	0.031	5.9	0.000***
Maximum log-likelihood: –155292.1				

Olive oil PDOs, fit for four latent classes:

2/1	Coefficient $\beta r,i$	Std. error	*t* value	$P (> \|t\|)$
Intercept	2.726	0.093	29.2	0.000***
Activities	−0.227	0.016	−14.1	0.000***
Shopping	−0.131	0.027	−4.8	0.000***
Goods & products	−0.066	0.003	−17.8	0.000***
Mobility	−0.425	0.022	−19.3	0.000***
3/1	Coefficient $\beta r,i$	Std. error	*t* value	$P (> \|t\|)$
Intercept	2.606	0.101	25.8	0.000***
Activities	−0.222	0.018	−12.0	0.000***
Shopping	−0.076	0.029	−2.6	0.000***
Goods & products	−0.060	0.004	12.1	0.000***
Mobility	−0.733	0.038	−19.2	0.000***
4/1	Coefficient $\beta r,i$	Std. error	*t* value	$P (> \|t\|)$
Intercept	2759	0.099	27.9	0.000***
Activities	−0.361	0.024	−15.2	0.000***
Shopping	−0.290	0.037	−7.8	0.000***
Goods & products	−0.091	0.005	−16.8	0.000***
Mobility	−0.651	0.037	−17.7	0.000***
Maximum log-likelihood: −146571.8				

Appendix B: Probability of Class Membership According to Each Analysed Expenditure

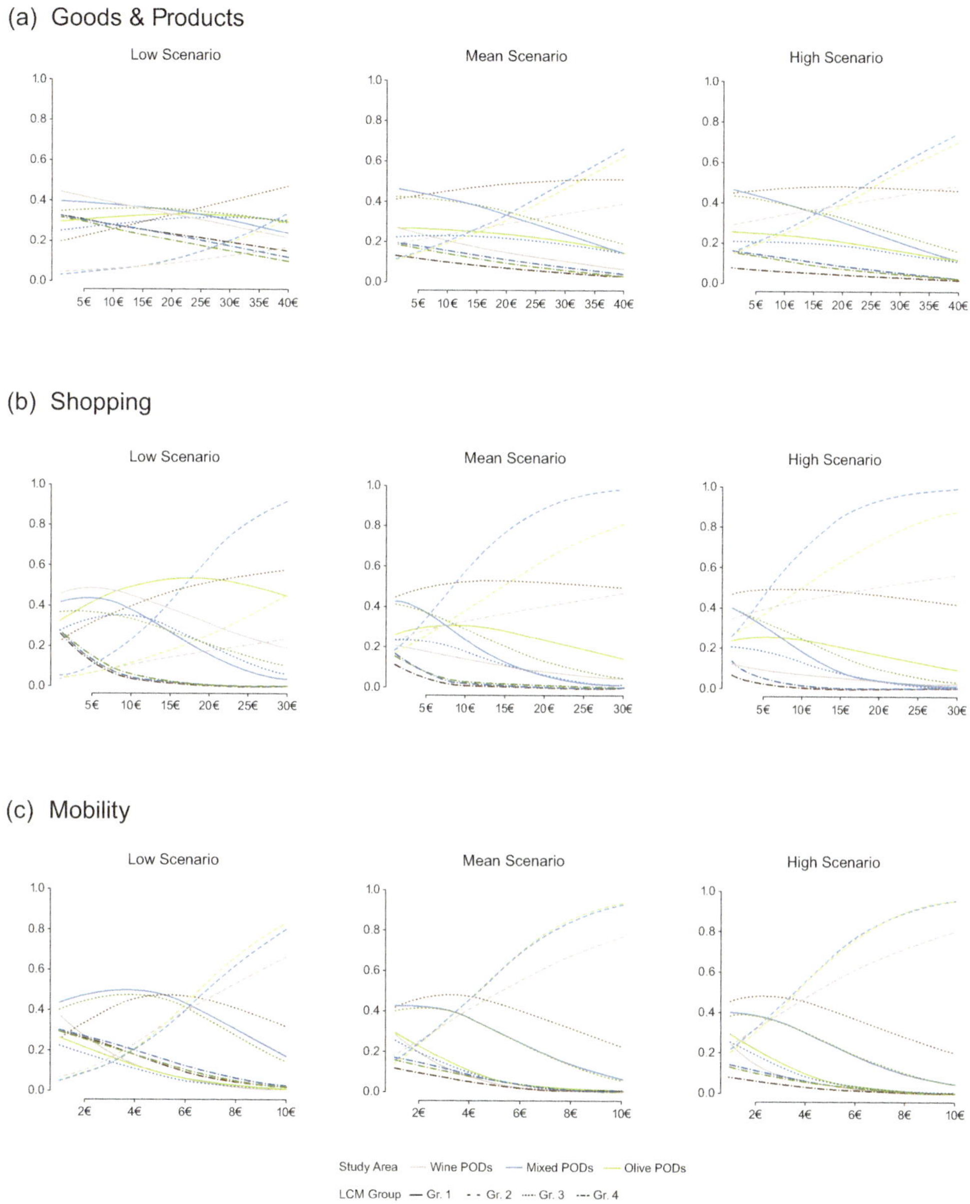

Fig. 5.9. Evolution of class membership according to three scenarios.

6 Managing Hybrid Destinations: Challenges and Lessons from the Alps

THOMAS BAUSCH*

Competence Centre Tourism and Mobility, Free University of Bozen, Bolzano, Italy

6.1 Introduction: The Need for Specific Marketing of Hybrid Destinations

In destination marketing literature the role of seasonality is a rarely discussed aspect. Pike and Page (2014) state in their review of areas of destination marketing, '...probably the most neglected area of DMO performance has been in relation to destination marketing and its impact on seasonality'. The academic literature database Scopus found only eight papers dealing with the keywords 'destination marketing' and 'seasonality', while nine are connecting 'destination development' and 'seasonality'. Those dealing with seasonality and tourism in most cases discuss destinations with one single high season (Turrión-Prats and Duro, 2018; Grossi and Mussini, 2021), very frequently from the Mediterranean. Thus, destinations with two or more tourist seasons, attracting their guests for each season by different types of services and related experiences, are not sufficiently covered by current research.

All regions where the climate causes a change of natural resources in such a way that the related outdoor activities swap every half-year and attract large but different markets fall into this type of destination. Although this change of the destination's characteristics is not discussed in the literature, it can frequently be found on the globe. Most mountain areas, which cover a large part of the world, are examples of this type of hybridity: in summer, the mild and moderate climate allows activities such as hiking, biking, climbing or just relaxing in nature, while in winter the profile changes towards winter sports and winter recreation. This hybrid situation calls for destination

*thomas.bausch@unibz.it

DOI: 10.1079/9781789248753.0006

marketing that considers the diverse seasonal product as well as seasonally changing target groups.

Also, the traditional concept of target groups as a result of market segmentation supposes that every consumer can be assigned to exactly one target group. Dolničar (2004) showed six different concepts for segmentation in tourism. All these concepts have an explanatory character (Dolničar, 2019) and are based on data describing the consumer as a static object because the data are static; however, humans are not static; they frequently change their preferences and their consumption choices, which was characterized by Ehrnrooth and Gronroos (2013) as 'hybrid consumption behavior'. This phenomenon can also be found in tourism. Destination marketing must consider that the guests frequently behave in a hybrid manner – they are hybrid tourists (Boztug *et al.*, 2015).

This chapter discusses hybridity of tourist destinations from a supply-and-demand perspective, using the example of the European Alps as the most frequently visited mountain destination in the world. Implications for destination marketing being faced by effects of hybridity are given.

6.2 The Alps as a Leading International Destination

The Alps are one of the most frequently visited destinations in the world. Although Europeans use the term 'Alps' as the name for a geographical area shared by seven countries (Austria, France, Germany, Italy, Liechtenstein, Monaco and Slovenia), there does not exist one single definition of what belongs to the Alpine territory. Depending on the perspective, which can differ significantly, the boundaries may vary considerably. Geologists look at the Alps as an area of 'folding' due to colliding tectonic plates, mainly the Eurasian plate and the Adriatic-African plate. According to them, everything that can be classified as the result of the folding process belongs to the Alps (Frisch, 1979). For the Alpine Convention, a transnational treaty for the protection of the Alps, the representatives of the aforementioned countries used administrative boundaries at municipality or district level in combination with the geological Alps to define the area (Permanent Secretary of the Alpine Convention, 2016).

The economic interdependencies as well as the emotional link between the population living in the Alpine mountain areas and those living in the nearby areas are traditionally strong. They need to be considered in strategic regional development plans (Gloersen *et al.*, 2013) as a development in the Alpine mountain areas interacts with the neighbouring areas. Therefore, from the perspective of the European regional development policy, the Alpine space was defined as a much larger area, adding the surrounding foothills and cities such as Lyon, Milan, Munich, Vienna and Zurich to the territory.

As there does not exist one common accepted definition of the Alpine territory, published statistics about Alpine tourism are fairly vague. In 2002, Pechlaner and Sauerwein (2002) reported 370 million overnight stays in 4.7

million guest beds for the Alps. WWF (Worldwide Fund for Nature) estimates the number of tourists visiting the Alps to be 120 million each year (WWF, n.d.). A Swiss benchmark study (BAKBASEL, 2014) published the market share of the Alpine tourism as 3.1% of the global tourism for 2012 with 496 million overnight stays in 7.5 million beds. Even though the numbers differ and are just rough estimations, these huge figures underline the fact that the Alps as a total area are a well-known and highly demanded destination in the global tourism industry.

Although the Alps as a whole are often seen as a unique destination, diverse characteristics require them to be divided into a larger number of sub-destinations:

- **Topography and geographic units**. Certainly, the mountains are the joint feature of the Alps; but in detail we find very different topographic situations. The geographic units are formed by the combination of mountain ranges, valleys with river basins and lakes. While the higher mountain ranges, in particular, offer tourism development options for the winter season, river basins and lakes attract more summer guests.
- **Settlement structure**. The majority of settlement areas are located in the valleys. Historically, they developed along transportation routes connecting the valleys via mountain passes that were traversable all year round. Since Roman times, the combination of passes and larger valleys has been considered as strategically important corridors across the Alps for trade and military purposes. Cities along the corridors, such as Bolzano, Innsbruck, Geneva, Grenoble, Klagenfurt, Ljubljana or Salzburg, therefore, have a long history. Nowadays, these cities are prosperous regional centres offering a broad spectrum of cultural attractions and services for leisure or business tourism.
- **Climate**. The Alps act as a natural barrier, separating the Mediterranean from the western and central European climate. The climate has a significant influence on the vegetation and, thus, on the agricultural and landscape structure. Whilst in the south agriculture is manifold, comprising the cultivation of all types of crops, vegetables and fruits, including grapes for viticulture, and allows all types of livestock breeding, the centre and north are mainly characterized by meadows and pastures offering limited cultivation options, mainly milk production.
- **Natural resources**. The variety and richness of local Alpine-specific natural resources are the result of the young Alpine geology. These resources are the basis for health and spa tourism: thermal and mineral springs, minerals from the rocks, natural ultra-violet or low nuclear radiation, or pure mountain air having health benefits in combination with the solar radiation. Therefore, looking at Alpine tourism historically, health and cure tourism of the 19th century had been an important development driver.
- **Cultural heritage and diversity**. The Alps have always been a barrier and a border, but at the same time a connection of cultures. Along the

transit routes crossing the Alps, a lively exchange between cultures has led to a richness of cultural heritage as a mix from both sides of the Alpine arc. Moreover, in peripheral and enclosed side valleys and mountain areas, often locally unique cultures evolved and have been preserved over centuries, e.g. by local languages like Ladin or Rhaeto-Romanic, or specific cultivation techniques in agriculture such as water management in arid areas or the jointly used rough pastures, the Allmende (Roeder *et al.*, 2010). All these cultural elements support touristic development up to the present day and create stereotypes with regard to the Alpine destination image.

The specific natural climate and cultural conditions in combination with the accessibility of each Alpine region induced the development of different types of tourism. The fourth report on the status of the Alps (Permanent Secretary of the Alpine Convention, 2013) lists seven main destination types: foothill summer, health, nature experience, Alpine cities and villages, all-season inner mountain, ski resorts and all-season destinations at the edge of the Alps. Therefore the Alps are a kind of destination jigsaw puzzle. On the one hand they are seen as one leading global tourism area with the main feature of high mountains, but on the other they are highly fragmented and diverse. We find a hybrid supply structure offered by a large number of destinations of high diversity.

6.3 Destination Image of the Alps and Its Brand Core Values

Destination image (Baloglu and McCleary, 1999) is individual for each traveller. In a qualitative study, Germans from different regions, of different age, gender and educational level were asked (Goetz *et al.*, 2011) what comes to their mind when thinking of the Alps. The results showed that participants reported a broad variety of personal destination images. These images are a mixture of cognitive and affective associations and frequently use the above-mentioned stereotypes of symbolic character. They are based on experiences that form autonomous image components as well as on other information sources, forming induced and organic image components (Tasci and Gartner, 2007; Kislali *et al.*, 2016). Not surprisingly, the mountains are the dominant image-forming element of the Alps. Nonetheless, people interconnect these mountains of very different dimensions at both cognitive and affective level. Hence, the core of the Alpine image, the mountains, creates very different connotations and is individually connected with different dimensions such as activities, places or emotions (see Table 6.1). Even though everybody would agree that the Alps are formed by mountains, the personal image is unique and can differ between travellers visiting exactly the same place. Descriptive attributes can lie on opposite sides of the same scale, e.g. happiness – panic; pure intact nature – erosion/infrastructure; grace and beauty of mountains – natural hazards; winter sports – summer activities.

Table 6.1. Image components of the Alps – results from qualitative study with 23 participants. (From Goetz *et al.*, 2011.)

Dimension	Associations
The ideal	blue sky, sun, fresh air glittering snow, glaciers, cold cows and alpine pastures mountain lakes, marmots alpine flowers, mountain meadows pure intact nature
Sports and activities	skiing, hiking, climbing
Aesthetics linked to nature	grace and beauty of mountains silhouette, monument of mountain mass, power of mountains
Alpine culture – authenticity	Cowbells, balconies with cranesbills friendly, grounded people traditional and delicious cooking idioms and ancient customs Heidi, Lederhosen, Dirndl
Essential emotions	freedom and independence endeavour, sweat and happiness breathlessness, panic standing on top of a summit with unique panoramic view unbelievably huge, wide, strong
Love and respect	loving mountains and mountain nature respect for forces of nature being aware of risks from elements
Regions and places	Tyrol, Ticino, Zugspitze, Matterhorn, Lake Constance, Ischgl, Garmisch, St Moritz, royal castles, forts/fortresses
Threats and negative development	erosion, climate change avalanches, flooding, rock falls cable cars, ski slopes, tourism centres

Considering the Alps not only as a regional area but also as a global brand, the question about the core brand values comes up. Pechlaner *et al.* (2007) showed that the mountains as a symbol for pure but also harsh nature are the key assets. This core value is used by companies from many economic sectors to increase the brand equity of their products: food producers, breweries and mineral water bottlers, car manufacturers, apparel industry, companies producing pharmaceuticals and cosmetics. Some of these products have their origin in the Alps; but in many cases the Alps and related stereotypes are just used as symbols to underline the purity and effectiveness of the products. The Alps, therefore, on the one hand, are a highly symbolic place umbrella brand (Iversen and Hem, 2008) but on the other, they are a geographical area with a large set of highly diverse destinations (Zehrer *et al.*, 2007).

The main difference between the Alps as an umbrella brand and corporate brands with a large portfolio of sub-brands, e.g. Unilever or Procter & Gamble (Gabrielli and Baghi, 2020), is the non-controllable brand-image formation. In the case of tourism, the brand image of the destination Alps has to be seen as identical to the destination image. Everything related to the consumer's individual image of the Alps contributes to the brand's equity (Konecnik and Gartner, 2007). Therefore, the observed heterogeneity of consumers' image of the Alps, which links very different touristic and non-touristic associations and connotations with mountains, suggests that the brand 'Alps' itself has a hybrid character.

6.4 Hybridity of Alpine Destinations: Supply-side Development Perspective

Leiper (1990) explained tourism as a system that consists of source markets and destinations linked by transit routes which allow the flow to and from the destinations. Attractions are the 'pull factor' motivating people to visit a destination and creating the flow from source markets to the destination. Diversity, accessibility and amenity of a destination's attractions are the basis for destination development. Destinations with just one attraction and one type of guest are called 'uniform' or 'non-hybrid'. Destinations having two or more attractions pulling guests with different expectations are 'binary-' or 'multiple-hybrid'. In many cases, the reason for hybridity is that the destination characteristics are not static and vary between seasons and over time. Climate is one key factor for most types of tourism based on natural attractions such as the seaside, lakes and rivers, hills and mountains, forests, steppes, wetlands and deserts. Climate is not a constant and therefore the level of attractiveness of a natural resource is linked with the potential level of fulfilment of the intended activities of guests: skiing not only requires mountains but also snow; for sunbathing and swimming in the sea, sunshine and temperatures within a comfort level are desirable (Orosa *et al.*, 2014). In certain cases, climate can create a twofold situation in tourism. In Europe, sun and beach tourism clearly is the largest holiday tourism market and the Mediterranean seaside offers perfect conditions from May/June to September/October. In the remaining months, sun and beach tourism are not feasible. Inversely, for winter sports we find a similar situation because the reliability of snow is a key asset for winter destinations' competitiveness. In the long term, climate change might lead to a shift in destination attractivity. Giannakopoulos *et al.* (2009) project an improvement of tourism conditions in the Mediterranean for spring and autumn, but a degradation during high summer. For the winter season, the effects of climate change will lead to more frequent snow scarcity and shorter seasons for skiing resorts (Steiger *et al.*, 2019). This will affect especially mountain destinations of lower altitude such as mid-range mountains or the foothills of the Alps.

Moreover, cultural attractions or events linked to a certain date or period of the year can serve as pull factors in tourism. In this case such attractions are interrelated with the place's natural environment and the traditions of locals; they support and strengthen the destination image formation. These cultural elements or events complement the natural destination characteristics. Frequently, the date or periods of these attractors are linked to the seasons; for example Christmas markets with the Advent season (Marcher *et al.*, 2019), carnival linked with the end of the winter; or wine and fruit festivals with the end of the harvest in autumn (Yuan *et al.*, 2008). However, if cultural attractions or events are only artificially created, destinations risk adding destination image components that are detached from the core image values, destabilizing them. Often, this can be observed in the field of yearly-held sports events in disciplines which do not have any connection to the touristic product. Mega-events such as world cups in ski jumping, bobsledding, biathlon or speed-skating, which are used as positioning elements for winter sports destinations, must be seen very critically. These are all niche sports, mainly performed at a professional level, and the built infrastructure for competitions is not available for amateurs and guests. Despite extensive TV coverage during a competition, it is unlikely that television viewers will develop the desire to visit these places just to see the infrastructure (Hallmann *et al.*, 2014). For example, some destinations added the component 'world cup venue' to their image and are now well known for a certain sports event. However, the destination brand equity could not be increased. Instead, the destination name became strongly associated with a one- or two-day event but lost the association with the original key assets (Chalip *et al.*, 2003). Examples are the biathlon world cup venues Oberhof and Rupolding in Germany. Nowadays, both locations are very well known due to their two- or three-day media coverage on German TV. But looking at the official tourism statistics (BayLfS, 2020), there is no steady above-average growth of arrivals and overnight stays to be seen. Also, the event did not stimulate larger investments into new accommodation facilities and attractions. Obviously, the biathlon world cup event does not create any desire to visit the destinations during the year other than for the event. For that reason, the diversification of a destination image must consider the existing key values when adding artificial elements. Diversification strategies that create a hybrid or, worse, a fragmented destination image should be scrutinized critically.

The hybridity of the Alps' destination image can be observed similarly for most of the Alpine destinations. Most of the above-listed destination types have two high seasons, the summer and the winter, supplemented by two low seasons in spring and autumn. This is due to the Alpine climate. During the winter season, many Alpine destinations have their positioning core on winter sports, some even with a stronger focus on just Alpine skiing. In this case, the destination brand core values are reduced to two main assets: slopes for skiing and snow-covered landscapes. For young people these two assets are the basis for a further affective characteristic linked to many Alpine winter destinations: fun (Matzler and Siller, 2003). But in summer, none of these assets exists. In summer, the destination image changes fundamentally.

Matzler and Siller (2003) showed that during that time, nature and landscape are associated with the emotion of 'freedom' for these young people. Enlarging the study to all age groups for Germany, 'idyll' and 'silence' are further key attributes of the summer destination image of Alpine destinations (Pröbstl-Haider *et al.*, 2015). Thus, even though the geographic names of the destinations are the same all year round, the destination image changes completely between winter and summer seasons.

The strong focus of many winter destinations on alpine skiing and après-ski is accompanied by the infrastructure needed to compete against other major alpine skiing destinations. Investments in high-performant mountain railways, facilities for snow production, or spacious slopes, change the landscape. In the villages, après-ski bars, coffee bars and restaurants, shops for ski-sports equipment and related services, ski rental services and offices for ski schools dominate the townscape during the winter season. The competitiveness of the ski destinations in winter is a result of these facilities and the related services. While winter guests appreciate what they find, a large share of potential summer guests frequently dislike this winter-related infrastructure (Pütz *et al.*, 2011). This results in a dilemma for further destination development: to stay competitive in winter, innovations for skiing guests are needed. These innovations are heavily linked with the construction of new, or replacement of old, infrastructure. However, the more the destinations advance in competitiveness for the winter sports market, the more they move away from attractiveness to potential summer guests, as these guests are bothered by the infrastructure and the winter-dominated townscapes.

As the European market for Alpine skiing is stagnating, and in the long term significantly shrinking (Bausch *et al.*, 2017; Vanat, 2020), the competition among skiing destinations will become tougher. This leads to the strategic question for skiing destinations: should they expand their positioning with non-skiing activities to attract new guests? Whilst for the accommodation sector of the destinations the type of guest does not make a big difference, this is a crucial point for the ropeway companies: every skiing guest substituted by a non-skier causes a cut in their revenue during the winter season. This calls for a decision to be made between strengthening the summer season by transitioning to a more summer guest-compatible product, or to further support the Alpine skiing sector. Thus, further destination development must consider the hybrid character of the product in light of the conflicting seasons in the long term. Looking at Richard Butler's tourism area life-cycle (TALC) model (Butler, 2006), we can see a development curve over time, starting with the involvement and ending up with scenarios between rejuvenation and decline. This supposes that a tourism area follows such a development curve. But a further, and in literature so far not discussed, aspect of hybridity can occur: while the winter season reaches the critical threshold of capacity and tends to decline, thus needing a rejuvenation strategy, the summer season might still be developing or moving to the phase of consolidation. The application of Butler's TALC model underlines that a successful destination development requires integrated strategies that consider both sides of the hybrid product character.

6.5 Alpine Source Markets: The Multiple-hybrid Tourist

Travelling is part of peoples' consumption and is a good example of the fact that the role of consumption in today's society is no longer about just fulfilling basic needs (Ehrnrooth and Gronroos, 2013). For many people, travelling adds to their identity creation, self-image and self-esteem, even though the contribution does not last a long time (Gabriel, 2006). On one hand, hybrid consumers seek to maximize their personal customer benefit by hunting for bargains or buying products and services that are good value for money; and on the other, the exact same consumers occasionally purchase more expensive goods, attaching a high emotional importance to increasing their personal value as much as possible. This phenomenon described by Ehrnrooth and Gronroos (2013) is the reason for hybrid consumption and we can find it in the tourism industry too. The growth of low-cost and budget segments on the one hand and full-service and luxury offers on the other can be observed in all areas of tourism: accommodation (Pütz-Willems, 2007); transportation, especially the airline industry (Ramón-Rodríguez *et al.*, 2011); tour-operators; as well as at destination level.

Tourists' individual travel behaviour as part of their consumption can be classified as 'multiple-hybrid' as they show hybridity in two dimensions (Boztug *et al.*, 2015):

- **Variation of expenditure**. First, different behaviour among travellers can be observed. Some tend to a low variation of travel budget for all trips, others to a high variation with strongly differing budgets per trip. The hybrid consumption patterns, characterized through switching between bargain-hunting for low-cost and last-minute offers and high-price luxury trips, are typical of travellers with high budget variation.
- **Motivation**. The motivations for all trips of some travellers are always the same; for others, every single trip has different motivation patterns.

From the perspective of destinations, the hybridity of travellers can also be understood as being confronted with a very heterogeneous guest structure concerning their motivations and expectations, as well as the willingness to pay. Furthermore, the behaviour of guests during the stay can vary between a constant daily routine and spontaneous activity decisions. Those guests who like to maximize outstanding experiences but also look for relaxation and calmness often behave in a hybrid manner. They appreciate the possibility to select from a wide range of options offered by the destination to cover their multi-optional needs situationally or spontaneously (Weiermair *et al.*, 2007).

As an example to show the behaviour of hybrid travellers, we take a closer look at the German source market. The following analysis uses the data from the yearly-held German travel analysis 'Reiseanalyse' (FUR, 2020). In this survey, 7500 representative selected Germans report in detail all holiday trips of the previous year, expenditure per trip, their general travel motives and personal socio-demographic information. Furthermore, all trips are classified by the survey participants as either main vacation or additional

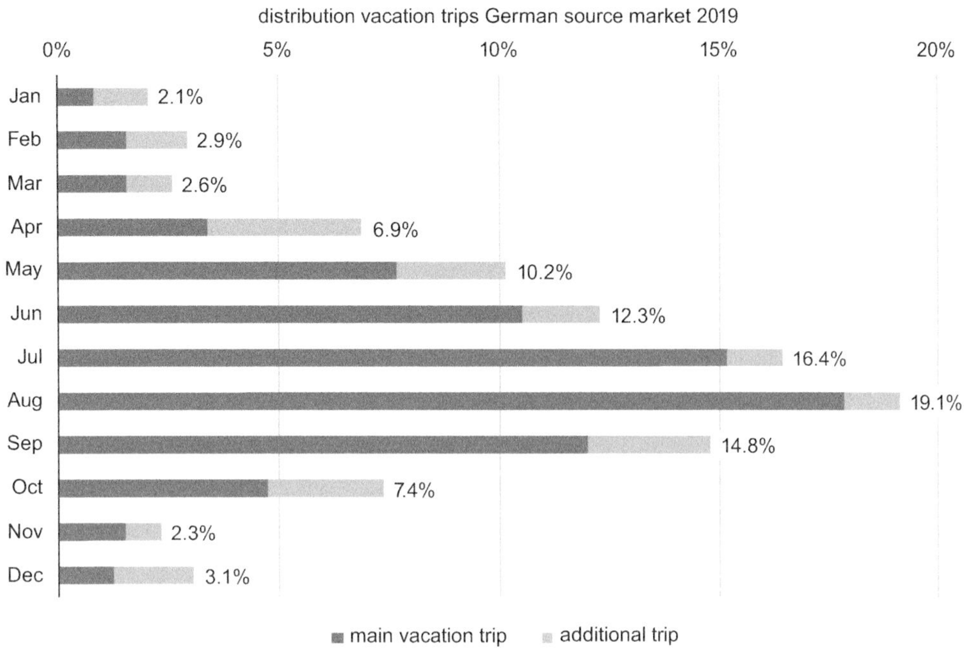

Fig. 6.1. Distribution of vacation trips for the German market.

vacation trips. Fig. 6.1 shows the seasonal distribution of vacation trips in 2019. The very unbalanced seasonal distribution of vacation trips shows that many more people travel in summer than in winter. From all vacation trips, 63% took place between June and September. By just looking at the main vacation trips, the imbalance grows even larger, as 71% of trips fall into the summer period. Apart from traditions and weather expectations, personal constraints, especially school holidays, university and company summer breaks, force people to travel during the summer period.

Guests from Germany who visited the Alps in 2019 for a minimum of five days showed substantial differences when comparing winter (January, February and December) and summer (June, July and August) guest characteristics (see Table 6.2). Winter guests are generally younger and have a higher share of male travellers in comparison to summer guests. On average, they have a higher education level which correlates with a higher income and expenditure for the Alpine winter holiday trips, even though the stays are shorter than in summer. The winter holiday trips are mostly classified as a holiday trip in addition to vacation trips during the summer. While Alpine summer guests rarely undertake an additional trip during the winter season, almost all of the winter guests go on a winter and a summer holiday trip. Thereby, the summer trip is classified as the most important, the main holiday trip. Both groups share one characteristic: they visit the Alps either in winter or in summer. People having both summer and winter vacations in the Alps are very rare. Therefore, Alpine holiday guests travelling more than once a

Table 6.2. Alpine winter and summer holiday guests' characteristics (source market: Germany, 2019). (Author's data analysis using results from German travel analysis.) (From FUR, 2020.)

Characteristic	Winter	Summer
Type of trip / duration of stay	secondary trips (86%)/8.7 days	main holidays (78%)/10.6 days
Expenditure per person and day / total travel budget (all persons)	104.22 € 1,949.83 €	78.12 € 1,558.07 €
Age	42.1 years average	53.1 years average
Gender (male : female)	62.1% : 37.9%	43.7% : 56.3%
Highest education level	30.7% high school; 27.2% university degree	20.3% high school; 11.0% university degree
Monthly net income of household	4,007.44 €	3,154.55 €
No. of holiday trips of five or more days	2.16	1.40
No. of short trips of two to four days	1.40	0.70
Further holiday trips in complementary season (summer for winter/ winter for summer)	have a main vacation in summer, but other destinations than Alps (92.5%); Mediterranean area (51%); northern seaside (14%)	only few with additional winter holiday trips (6%); only 2% with Alps as winter destination
Five most important travel motives	having fun (70%), distance from daily life (68%), having time for each other (66%), sun and warmth (66%), relaxation (64%)	distance from daily life (71%), nature experience (69%), regain energy (63%), having time for each other (62%), sun and warmth (58%)

year must be classified as 'destination hybrid': they change the type of destination between summer and winter. Half of the Alpine winter guests prefer to visit the Mediterranean in summer. Looking at the main travel motives of winter and summer guests we can also clearly identify major differences: winter guests expect fun while summer guests have a much stronger focus on nature experience, calmness and regaining energy.

Significant differences can be found not only in regard to the travel motives between Alpine winter and summer guests but also with regard to their expectations. Bausch and Gartner (2020) analysed the expectations of German travellers interested in a winter holiday trip to the Alps. Based on 20 items describing the expectations German travellers have of a winter holiday trip, which were derived in a qualitative study (Bausch and Unseld, 2018), they developed a typology of potential German winter guests. They found ten types of expectation profiles, varying from sport-focused groups expecting just snow and skiing, while other groups were expecting a broad variety of

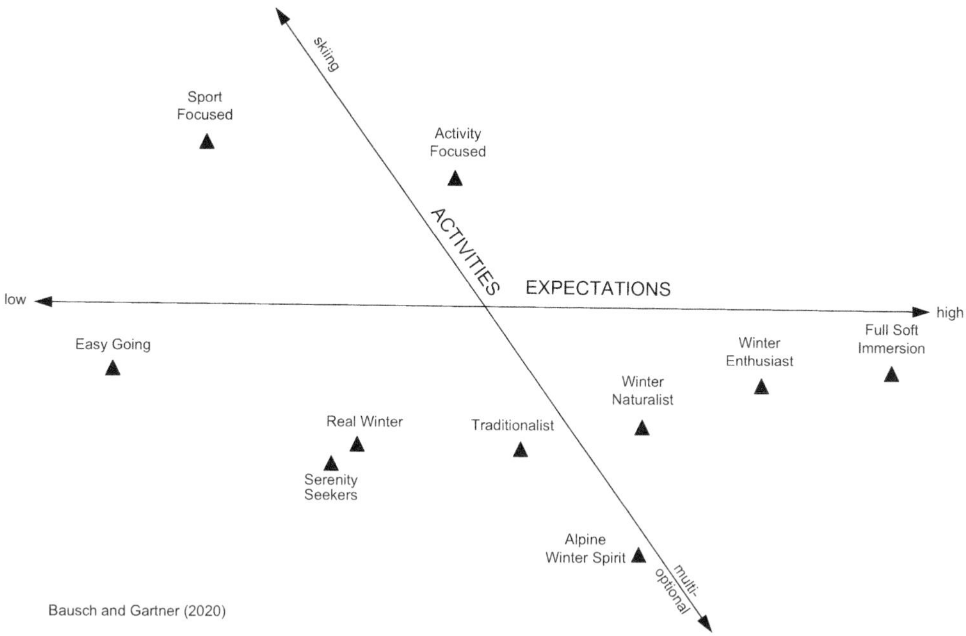

Fig. 6.2. Typology presentation of German winter tourists along axes of expectations and activities.

nature-based winter experiences, typical regional food and accommodation with a cosy atmosphere, but not, or only occasionally, skiing. The ten winter tourist types could be projected using multi-dimensional scaling along two axes (see Fig. 6.2): the expectation level (low–high) and the activities (just skiing–multi-optional activity setting). The study showed the hybridity of the German winter tourism market.

It became visible that most leading Alpine winter destinations potentially neglect guests who are not, or are only slightly, interested in skiing, although there is great potential and a great market share to skim. These groups have significantly varying interests: those with higher expectation levels share expectations in hiking in a snow-covered winter landscape and enjoyment of the Alpine cultural atmosphere in combination with food and beverages.

According to Fig. 6.2, most Alpine winter destinations try to attract the guest types 'sport-focused' and 'activity-focused'. This has led to extensive competition, requiring huge investments, e.g. 1.58 billion euros between 2015 and 2020 in northern Tyrol (Austria) (WKO, 2020). Falk and Steiger (2020) showed that the length of the slope network of a ski resort is the most important success factor. But in most Alpine skiing destinations, the slope network cannot be expanded any more due to physical or environmental limitations: most mountains suitable for installing cable cars and slopes are already in use or declared as protected areas. Therefore the strategy of many destinations was to connect neighbouring skiing resorts to enlarge the slope system concerning size, variability and share of slopes with reliable snow coverage

(Kuščer and Dwyer, 2019). This automatically is linked to the location of significant parts of the entire slope system in high altitude. However, this needs either interlinking of further ski resorts or, in rare cases, an expansion of ski resorts with new slopes at higher altitude. Further interlinkages will be very rare, as most options for creating larger ski areas are already realized. Equally, opportunities for new expansions and interlinkages at high altitude are very rare and cause heavy conflicts with environmental NGOs. Finally, such projects entail negative media coverage (Hruby, 2020) and underline the criticisms against Alpine skiing tourism. Generally, a further growth of ski resorts in the Alps is very limited. Therefore, the usual practice to improve competitiveness over the last decades was the replacement of older and smaller cable cars and stations with new and high-performing systems. These new systems need larger station buildings at both valley and mountain stations, as well as larger and higher pillars. Such infrastructure measures might be suitable to compete in the segment of Alpine winter sports guests, but at the same time they are highly adverse for improving competitiveness for nature-oriented summer and winter guests, as each infrastructure modernization means further deterioration of the landscape.

6.6 Conclusion and Implications

Using the example of the Alps, it was shown that hybridity in tourism touches primarily three marketing fields:

1. **Product**. Because of attractions that change their character and level of appeal throughout the year or are linked to yearly events and habits, the perception of the core elements of the destination's image varies.
2. **Communication**. The hybridity of the product implies the need to address different target groups, as the changing focus of the destination image attracts different, often disjunctive or mutually exclusive, consumer groups.
3. **Pricing**. Similar to all fields of consumption, an increasing polarization between low-cost or bargain trips and high-priced luxury travelling can be found. Some travellers prefer either low-cost or luxury travelling, while others change their expenditure from trip to trip.

The following discussion focuses on mature destinations having at least one high season with largely full occupancy.

The management options in the field of product policy depend on the type of hybridity of the destination, which were described above as: uniform non-hybrid, binary-hybrid and multiple-hybrid. The climate there is the most important factor for uniform and binary-hybrid vacation destinations. However, climate is a given locational factor and cannot be influenced. Looking at the Mediterranean, historically, the majority of destinations developed along the coastline with large hotels, campgrounds or resorts with holiday homes having the beach in front of them. The hinterland of these destinations, with its attractions, did not play any role in the development phase of the tourism area life-cycle in most cases (Tyrakowski, 1986). The destinations were designed as uniform non-hybrid products attracting non-hybrid guests with very uniform motives: sun and beach. Therefore, the role of destination management, in many cases, is just a management of peak effects during the high season. Communication and pricing are largely in the hands of the suppliers or their co-operating tour operators (Rigall-I-Torrent and Fluvià, 2011). In many cases, the hotels are owned by large groups, some being totally or partially owned by a tour operator, such as Iberotel (100%) or RIU-hotels (49%) in the TUI-group (TUI Group, 2021). The key problems for this type of destination for a further development are twofold; first, the short summer season; and second, independent of season, the limited access to capacities and management of strategically outstanding accommodation by the destination management caused by the described tenure structure. During the high season, with its midsummer weather, a diversification or 'hybridization' of the product is doomed to failure, as the majority of guests is not interested in discovering the hinterland. There is no real market for new, innovative other products. Additionally, even in case that innovation could be an instrument for rejuvenation, the availability of further resources along the coastline has been very limited for more than three decades (Tyrakowski, 1986). Furthermore, in the pre-Covid-19 era, the discussion about overtourism had led to the question of a need for degrowth (Valdivielso and Moranta, 2019).

Thus the idea of a future sustainable development of such destinations is a degrowth of high season in combination with a stepwise development of the side seasons. Looking at the seasonal distribution of vacation trips in Europe (compare Fig. 6.1 for Germany), this seems to be a fight against the rules of the market and its traditions. In Mediterranean countries, escaping the extreme heat in cities during high season is a further aspect limiting the shift of holidays to the side seasons. April, May and October, nevertheless, are quite attractive travel months and the relatively high and increasing share of European households without children opens options for many people to move the vacation from the mid-season to the side.

This time shift, however, will not be performed by the core group of guests with the uniform profile of sun and beach. More likely, travellers who like to combine some hours at the beach with other activities such as hiking, biking or visiting cultural attractions, will be shifting their holidays and should therefore be addressed by the destination management. To do so, new types of co-operation between the accommodation sector at the coast and the destination management are needed, integrating attractions from the hinterland or developing new activity options. As good-practice examples, the step-by-step development of hiking and biking tourism on the islands of Crete, Mallorca or Cyprus are frequently mentioned (Rejón-Guardia *et al.*, 2018). However, these are small niche markets and only those destinations that are able to offer a good network of hiking or biking courses in an attractive nature will be recognized as destination choice options.

Binary hybrid product profiles can be observed, especially for mature destinations with a strong winter and summer season. As described above, the guests' expectations and profiles of winter and summer are quite different. At first glance, this seems to be a very comfortable situation: in winter the destination concentrates on a sufficiently large and profitable target group, and in summer on a disjunctive other one. The crucial question here is whether product development and the related impact on the destination image can be done independently for both seasons. Looking at the destination image formation process, as described by Tasci and Gartner (2007), this would either require that winter and summer image are not contradicting each other or that the addressed seasonal target groups are not influenced by the image of the other season. The first condition of not contradicting images assumes to attract both seasonal target groups keeping the promise to fulfil their travel motives and related expectations. Nevertheless, in the case where the one guest profile is based on fun and modern powerful technical infrastructure, and the other guest group expects calmness and pure, untouched nature, this will not work: the infrastructure needed to satisfy the one group is visible from afar and has a negative impact on the landscape perception of the other group. As for the German source market, the three strongest associations with a holiday in the Alpine area are: natural and undisturbed landscapes; idyll; and silence (Pröbstl-Haider *et al.*, 2015). This problem exists for all winter destinations with plenty of visible infrastructure.

One strategy of highly competitive winter sports destinations that can be found frequently is to target the same group of younger winter sports enthusiasts also during summer. Instead of skiing, they offer trendy sports such as downhill mountain-biking, trail-biking, uphill e-biking, paragliding or mountain-running. In addition, adventure parks, high-ropes courses, climbing and boulder walls, or bike parks, as further infrastructure as well as thematic events, are aimed at attracting this target group. Extensive advertisement during the ski season, promoting these summer activites will persuade winter guests to visit also in summer. However, this strategy works only in a very few cases and does not at all create the same guest flow as in winter. As an example, the destination Ötztal in Tyrol (Austria) may be taken. The village Sölden, as part of this destination, is one of the most competitive Tyrolean winter destinations with 2.59 million overnight stays in 2019, 80% of them during the winter season. Despite the creation of new infrastructure for summer activities, e.g. the largest leisure park of Austria, called AREA 47 (see www.area47.at), the region did not achieve significant change in the winter–summer guest ratio. Further similar examples are Saalbach (Austria), Flims-Laax and Verbier (Switzerland), Chamonix (France) and Bormio (Italy) trying to strengthen the summer season by addressing young sporting guests. One reason for the limited success is that the highly sporty guest segment is relatively small. The second is that the majority of young travellers prefer the seaside for their summer holidays. Thus, in the long term, binary-hybrid destinations will always have to choose between two options: either to maximize the competitiveness and economic benefit of one season and to live with the fact that the other season will never perform at the same level, or to adapt destructive image elements from both seasons to create a balanced system. It seems likely that such a balanced system will not compete with the specialized destinations, for guests having high expectations for just one activity. On the other hand, it will be able to attract large target groups wishing to experience a bit of everything. These potential guests are neglected by the specialists because of their focus on just one or a few attractions. In the long run this might be a sustainable approach as it is more resilient against mega-trends such as climate change, demographic change and disruption caused by regional or global crises.

Multiple-hybrid destinations are often the result of merging originally smaller destinations to one larger destination unit. Bieger *et al.* (2009) state that the destination size is a relevant factor for destinations to compete in the market. The authors argue that the extension of tourism destination boundaries is needed to turn the focus of destination management organizations from product orientation towards market-oriented service centres. Concentrating on the effective management of service processes, the extension of destination size is, for sure, an advantage. But it is also very likely that the level of hybridity of the destination increases as each further complemented area introduces additional attractions as new elements of the destination image. This finally entails the risk to lose a clear destination profile and to end up as a mish-mash. The destination can offer a little bit of everything, but there will always be competitors who are better in offering specific elements. Nevertheless, multiple-hybrid destinations are often successful as they are able to satisfy the expectations of the multiple-hybrid guests who wish for a tourism product where they can choose from a large set of options. As the single attractions are often not unique in these destinations, good accessibility and infrastructure, especially for accommodation, is a prerequisite for competitiveness.

References

BAKBASEL (2014) Benchmarking Turismo – Il settore turistico svizzero nel confronto internazionale. Rapporto sul 'Programma di benchmarking internazionale per il turismo svizzero: fase di progetto 2012–2013'. Basel. Available at: https://rea.bak-economics.com/fileadmin/documentsREA/BAKBASEL_Tourismus_Benchmarking_it_ready_def.pdf (accessed 24 June 2021).

Baloglu, S. and McCleary, K.W. (1999) A model of destination image formation. *Annals of Tourism Research* 26(4), 868–897. DOI: 10.1016/S0160-7383(99)00030-4.

Bausch, T. and Gartner, W.C. (2020) Winter tourism in the European Alps: is a new paradigm needed? *Journal of Outdoor Recreation and Tourism* 31(2), 100297. DOI: 10.1016/j.jort.2020.100297.

Bausch, T. and Unseld, C. (2018) Winter tourism in Germany is much more than skiing! Consumer motives and implications to Alpine destination marketing. *Journal of Vacation Marketing* 24(3), 203–217. DOI: 10.1177/1356766717691806.

Bausch, T., Ludwigs, R. and S, M. (2017) Winter tourism and climate change: impacts and adaptation strategies. Munich University of Applied Sciences, Department of Tourism. Available at: https://www.researchgate.net/publication/313892568_Winter_Tourism_and_Climate_Change_-_Impacts_and_adaptation_strategies (accessed 5 March 2017).

BayLfS (2020) Bavarian State Agency for Statsitics, monthly tourism data (cade 45511). © Bayerisches Landesamt für Statistik. Available at: https://www.statistikdaten.bayern.de/genesis/online/data?operation=statistic&levelindex=0&levelid=1596692334152&code=45511 (accessed 6 August 2020).

Bieger, T., Beritelli, P. and Laesser, C. (2009) Size matters! Increasing DMO effectiveness and extending tourism destination boundaries. *Tourism (13327461)* 57, 309–327.

Boztug, Y., Babakhani, N., Laesser, C. and Dolničar, S. (2015) The hybrid tourist. *Annals of Tourism Research* 54(4), 190–203. DOI: 10.1016/j.annals.2015.07.006.

Butler, R. (ed.) (2006) *The Tourism Area Life Cycle. Aspects of Tourism 28.* Channel View Publications, Buffalo, New York and Clevedon, UK.

Chalip, L., Green, B.C. and Hill, B. (2003) Effects of sport event media on destination image and intention to visit. *Journal of Sport Management* 17(3), 214–234. DOI: 10.1123/jsm.17.3.214.

Dolničar, S. (2004) Beyond 'commonsense segmentation': a systematics of segmentation approach in tourism. *Journal of Travel Research* 42(3), 244–250. DOI: 10.1177/0047287503258830.

Dolničar, S. (2019) Market segmentation analysis in tourism: a perspective paper. *Tourism Review* 75(1), 45–48. DOI: 10.1108/TR-02-2019-0041.

Ehrnrooth, H. and Gronroos, C. (2013) The hybrid consumer: exploring hybrid consumption behaviour. *Management Decision* 51. *Emerald Group Publishing Limited*, 1793–1820.

Falk, M. and Steiger, R. (2020) Size facilitates profitable ski lift operations. *Tourism Economics* 26(7), 1197–1211. DOI: 10.1177/1354816619868117.

Frisch, W. (1979) Tectonic progradation and plate tectonic evolution of the Alps. *Tectonophysics* 60(3-4), 121–139. DOI: 10.1016/0040-1951(79)90155-0.

FUR (2020) Dataset from German Travel Analysis 2020 provided by the research group vacation and travelling (FUR). Kiel. Available at: www.reiseanalyse.de (accessed 24 June 2021).

Gabriel, Y. (2006) *The Unmanageable Consumer*, 2nd edn. Sage Publications, Los Angeles, California.

Gabrielli, V. and Baghi, I. (2020) Unveiling the corporate brand: the role of portfolio composition. *Journal of Consumer Marketing* 37(3), 279–290. DOI: 10.1108/JCM-12-2018-2995.

Giannakopoulos, C., Le Sager, P., Bindi, M., Moriondo, M., Kostopoulou, E. *et al.* (2009) Climatic changes and associated impacts in the Mediterranean resulting from a 2°C global warming. *Global and Planetary Change* 68(3), 209–224. DOI: 10.1016/j.gloplacha.2009.06.001.

Gloersen, E., Bausch, T., Hurel, H., Pfefferkorn, W., Dal Fiore, F. *et al.* (2013) Strategy Development for the Alpine Space. Final Report (ed. Management Authority Alpine Space Programme). Salzburg. Available at: http://www.alpine-space.org/2007-2013/fileadmin/media/Downloads_in_about_the_programme/SDP_Final_Report.pdf (accessed 5 March 2017).

Goetz, K., Birzle-Harder, B. and Bausch, T. (2011) Lifestyles, holliday orientations and travel behaviour considering climate- and demographic change: the example of the Alps [Lebensstile, Urlaubsorientierungen und Reiseverhalten unter dem Aspekt von Klima-und demographischem Wandel am Beispiel der Alpen]. Munich / Frankfurt a.M. Available at: https://www.researchgate.net/publication/348651213_Lebensstile_Urlaubsorientierungen_und_Reiseverhalten_unter_dem_Aspekt_von_Klima-und_demographischem_Wandel_am_Beispiel_der_Alpen_Bericht_zu_einer_qualitativen_empirischen_Studie_im_Auftrag_der_Hochsch/related (accessed 24 June 2021).

Grossi, L. and Mussini, M. (2021) Seasonality in tourist flows: decomposing and testing changes in seasonal concentration. *Tourism Management* 84(355), 104289. DOI: 10.1016/j.tourman.2021.104289.

Hallmann, K., Müller, S. and Feiler, S. (2014) Destination competitiveness of winter sport resorts in the Alps: how sport tourists perceive destinations? *Current Issues in Tourism* 17(4), 327–349. DOI: 10.1080/13683500.2012.720247.

Hruby, D. (2020) In the Alps, ski resorts are desperately battling climate change – and local resistance. National Geographic. Available at: https://www.msn.com/en-us/weather/topstories/in-the-alps-ski-resorts-are-desperately-battling-climate-change-and-local-resistance/ar-BB1caNAm (accessed 17 February 2021).

Iversen, N.M. and Hem, L.E. (2008) Provenance associations as core values of place umbrella brands: a framework of characteristics. *European Journal of Marketing* 42, 603–626.

Kislali, H., Kavaratzis, M. and Saren, M. (2016) Rethinking destination image formation. *International Journal of Culture, Tourism and Hospitality Research* 10(1), 70–80. DOI: 10.1108/IJCTHR-05-2015-0050.

Konecnik, M. and Gartner, W.C. (2007) Customer-based brand equity for a destination. *Annals of Tourism Research* 34(2), 400–421. DOI: 10.1016/j.annals.2006.10.005.

Kuščer, K. and Dwyer, L. (2019) Determinants of sustainability of ski resorts: do size and altitude matter? *European Sport Management Quarterly* 19(4), 539–559. DOI: 10.1080/16184742.2018.1550097.

Leiper, N. (1990) Tourist attraction systems. *Annals of Tourism Research* 17(3), 367–384. DOI: 10.1016/0160-7383(90)90004-B.

Marcher, A., Erschbamer, G. and Pechlaner, H. (2019) Escaping from the event? Residents' perception of Christmas markets in Northern Italy. *Event Management* 23(4), 599–611. DOI: 10.3727/152599519X15506259855715.

Matzler, K. and Siller, H.J. (2003) Linking travel motivations with perceptions of destinations: the case of youth travelers in Alpine summer and winter tourism. *Tourism Review* 58(4), 6–11. DOI: 10.1108/eb058418.

Orosa, J.A., Costa, Á.M., Rodríguez-Fernández, Á. and Roshan, G. (2014) Effect of climate change on outdoor thermal comfort in humid climates. *Journal of Environmental Health Science and Engineering* 12(1), 46. DOI: 10.1186/2052-336X-12-46.

Pechlaner, H. and Sauerwein, E. (2002) Strategy implementation in the Alpine tourism industry. *International Journal of Contemporary Hospitality Management* 14(4), 157–168. DOI: 10.1108/09596110210427003.

Pechlaner, H., Raich, F. and Zehrer, A. (2007) The Alps: challenges and potentials of a brand management. *Tourism Analysis* 12(5), 359–369. DOI: 10.3727/108354207783228000.

Permanent Secretary of the Alpine Convention (2013) RSA4: Sustainable Tourism in the Alps. Innsbruck Available at: https://issuu.com/alpconv/docs/rsa4_en_full/51 (accessed 18 January 2021).

Permanent Secretary of the Alpine Convention (2016) The Alps – eight countries, one territory. Available at: https://www.alpconv.org/en/home/news-publications/publications-multimedia/detail/the-alps-eight-countries-one-territory-1/ (accessed 13 January 2021).

Pike, S. and Page, S.J. (2014) Destination Marketing Organizations and destination marketing: a narrative analysis of the literature. *Tourism Management* 41(3), 202–227. DOI: 10.1016/j.tourman.2013.09.009.

Pröbstl-Haider, U., Haider, W., Wirth, V. and Beardmore, B. (2015) Will climate change increase the attractiveness of summer destinations in the European Alps? A survey of German tourists. *Journal of Outdoor Recreation and Tourism* 11(4), 44–57. DOI: 10.1016/j.jort.2015.07.003.

Pütz, M., Gallati, D., Kytzia, S., Elsasser, H., Lardelli, C. *et al.* (2011) Winter tourism, climate change, and snowmaking in the Swiss Alps: tourists' attitudes and regional economic impacts. *Mountain Research and Development* 31(4), 357–362. DOI: 10.1659/MRD-JOURNAL-D-11-00039.1.

Pütz-Willems, M. (2007) Polarization of markets: luxury and budget hotels. In: Conrady, P.D.R. and Buck, D.M. (eds) *Trends and Issues in Global Tourism 2007*. Springer, Berlin Heidelberg, pp. 33–40.

Ramón-Rodríguez, A.B., Moreno-Izquierdo, L. and Perles-Ribes, J.F. (2011) Growth and internationalisation strategies in the airline industry. *Journal of Air Transport Management* 17(2), 110–115. DOI: 10.1016/j.jairtraman.2010.11.002.

Rejón-Guardia, F., García-Sastre, M.A. and Alemany-Hormaeche, M. (2018) Motivation-based behaviour and latent class segmentation of cycling tourists: a study of the Balearic Islands. *Tourism Economics* 24, 204–217.

Rigall-I-Torrent, R. and Fluvi, M. (2011) Managing tourism products and destinations embedding public good components: a hedonic approach. *Tourism Management* 32(2), 244–255. DOI: 10.1016/j.tourman.2009.12.009.

Roeder, N., Lederbogen, D., Trautner, J., Bergamini, A., Stofer, S. *et al.* (2010) The impact of changing agricultural policies on jointly used rough pastures in the Bavarian Pre-Alps: an economic and ecological scenario approach. *Ecological Economics* 69(12), 2435–2447. DOI: 10.1016/j.ecolecon.2010.07.013.

Steiger, R., Scott, D., Abegg, B., Pons, M. and Aall, C. (2019) A critical review of climate change risk for ski tourism. *Current Issues in Tourism* 22(11), 1343–1379. DOI: 10.1080/13683500.2017.1410110.

Tasci, A.D.A. and Gartner, W.C. (2007) Destination image and its functional relationships. *Journal of Travel Research* 45(4), 413–425. DOI: 10.1177/0047287507299569.

TUI Group (2021) TUI Group Shareholdings. Brighton, UK. Available at: https://www.tuigroup.com/en-en/investors/corporate-governance/list-of-shareholdings (accessed 16 February 2021).

Turrión-Prats, J. and Duro, J.A. (2018) Tourist seasonality and the role of markets. *Journal of Destination Marketing & Management* 8(2), 23–31. DOI: 10.1016/j.jdmm.2016.11.004.

Tyrakowski, K. (1986) The role of tourism in land utilization conflicts on the Spanish Mediterranean coast. *GeoJournal* 13(1), 19–26. DOI: 10.1007/BF00190685.

Valdivielso, J. and Moranta, J. (2019) The social construction of the tourism degrowth discourse in the Balearic Islands. *Journal of Sustainable Tourism* 27(12), 1876–1892. DOI: 10.1080/09669582.2019.1660670.

Vanat, L. (2020) 2020 International Report on Snow & Mountain Tourism. 12th edn, April. Geneva. Available at: https://www.vanat.ch/RM-world-report-2020.pdf (accessed 9 February 2020).

Weiermair, K., Peters, M. and Schuckert, M. (2007) Destination development and the tourist life-cycle: implications for entrepreneurship in Alpine tourism. *Tourism Recreation Research* 32(1), 83–93. DOI: 10.1080/02508281.2007.11081526.

WKO (2020) Tyrolean cable car economy in numbers (2019) [Tiroler Seilbahnwirtschaft in Zahlen 2019]. June. Innsbruck. Available at: https://www.wko.at/branchen/t/transport-verkehr/seilbahnen/FS_Seilbahnen2019.pdf (accessed 12 February 2021).

WWF (n.d.) Alpine Tourism. Available at: https://wwf.panda.org/discover/knowledge_hub/where_we_work/alps/problems/tourism/ (accessed 13 January 2021).

Yuan, J., Morrison, A.M., Cai, L.A. and Linton, S. (2008) A model of wine tourist behaviour: a festival approach. *International Journal of Tourism Research* 10(3), 207–219. DOI: 10.1002/jtr.651.

Zehrer, A., Pechlaner, H. and Raich, F. (2007) Destination Alps and its communicated brand image. *Anatolia* 18(2), 319–333. DOI: 10.1080/13032917.2007.9687208.

7 Diversifying and Enhancing Mountainous Destinations through Rural Tourism: The Case of Kalavrita, Greece

Nikolaos Boukas*

European University Cyprus, Cyprus

7.1 Introduction

While Greece is a popular destination for its many islands suitable for summer vacations, its physiognomy comprises also, to a large degree, continental places and many mountains. Nonetheless, these places have not been equally treated by tourism marketing as significant destinations that could generate volumes of international and domestic tourists and for this reason remain to a large degree unutilized. As such, many opportunities for Greek continental tourism as well as the appreciation of its value for these areas are lost. As a result, the tourism landscape of the country is highly seasonal concentrating only on certain mass-tourism products on specific months (mainly summer) and on particular destinations, while other regions of the country, such as mountainous regions, not only do not enjoy the same benefits but also face serious problems of decentralization and economic disparity. In this regard, tourism in these areas, on the one hand, would spread benefits to the country in a more balanced way and, on the other, would help Greece to diversify its tourism product/service mix as a whole-year destination.

Greece – facing, to a large degree, the consequences of mass tourism (especially in its islands), which have been exacerbated by the economic crisis (Pappas and Papatheodorou, 2017; Tsilimigkas and Derdemezi, 2020) – seeks to reconfigure and diversify its tourism product as a remedy to increase competitiveness of the tourism sector and follow a more sustainable direction (Stylidis and Terzidou, 2014; Stratigea and Katsoni, 2015) in its tourism development. Towards this end, the purpose of this chapter is to examine the

*n.boukas@euc.ac.cy

DOI: 10.1079/9781789248753.0007

case of the Kalavrita mountain region, located on Northern Peloponnesus, Greece, and the planning and marketing efforts that national and regional tourism authorities undertake to foster rural tourism development. The chapter emphasizes the tourism products that the destination offers, highlighting rural tourism as a macro-niche that is worthy to be considered as a unique form of tourism, and suggests a series of strategies for the diversification and enhancement of the tourism offering for the holistic Greek destination experience. In this regard, this study was employed for: (i) discovering the role of rural tourism for the mountainous area of Kalavrita, highlighting the challenges and opportunities of the current tourism situation in the area; (ii) analysing the strategies formulated by tourism authorities to engage in rural tourism; and (iii) conceptualizing the current and potential rural tourism products that do, and could, lead to the appropriate diversification of the existing tourism product of the Kalavrita region, as a representative western European mountainous destination.

7.2 Rural Tourism: The Need for Tourism Product Diversification

A very important aspect when considering tourism marketing is product diversification. Diversification is a strategy that can lead to the creation and maintenance of competitive advantage in the marketplace (Li and Greenwood, 2004). As Benur and Bramwell (2015) indicate, diversification, intensification and linkage of primary tourism products are significant for a destination to reach its competitive edge, as well as its sustainability. Rural tourism could be this form of tourism development where, on the one hand, strong backward linkages with other industries (i.e. agriculture) can be met in a very engaging manner (leading to more sustainable development), and, on the other hand, innovative products and processes met in this type of tourism development can lead to the appropriate diversification of a destination's tourism product and hence to its stronger position within competition. According to Hjalager (1996, p. 103), rural tourism contributes to the innovation of an area's tourism product as it is of small scale and highly associated with green issues, while its special facilities can differentiate significantly the product from others. As Hjalager *et al.* (2018) claim, rural tourism is characterized by both 'backwardness' in terms of history, authenticity and nostalgia for the traditional way of living and working, but also modernity in terms of rural processes such as novelty, comfort and re-inventions of traditional rural activities. This transformation and connection between old and new could be a desirable blend for a mountainous area's product diversification.

Nonetheless, one of the challenges of rural tourism is the full appreciation of its character. The challenge lies in the fact that the term 'rural' may be interpreted in different ways when examining different contexts (Hall *et al.*, 2003). Indeed, rurality has to do with the countryside. Nonetheless, rural areas can have multiple layers, are characterized by dynamism, are highly influenced by globalization, and are deeply connected to urban patterns and characteristics (Lane, 1994; Daspher, 2014). This means that the features of rural tourism evolve as the rest of the world does. As such, rurality connected to the traditional way of life meets modernity connected to the contemporary one.

In this respect, Lane (1994, p. 14) argues that rural destinations are those with the following conditions: they are located in rural areas; they are rural regarding their function (i.e. small-scale, traditional, entrepreneurship-focused and highly linked to the natural environment); they are of small scale; they are linked to family control and operations; and they are characterized by diversity, signifying the density of rurality in socioeconomic and cultural life. All those conditions develop a macro-niche (Robinson and Novelli, 2005) that incorporates various types of tourism (micro-niches) such as touring water-related activities, cultural tourism, religious tourism, ecotourism, event tourism, sporting activities, business tourism, health tourism, agricultural tourism, and so on, all taking place in the rural environment, linked to primary-sector activities, and, by definition, considered as more sustainable in terms of environment, culture, society and economy for the local destination (Roberts and Hall, 2001; Gao and Wu, 2017; Boukas, 2019). As such, rural tourism is not considered a single tourism activity, but rather a spectrum of tourism activities that take place in areas characterized by rurality, and also dynamism and evolution, in sociocultural and economic terms.

Unquestionably, therefore, rural tourism can be an appropriate form of tourism for rural areas such as Kalavrita and its significance is multifaceted (Egbali and Nosrat, 2011; Hernández *et al.*, 2016; Rasoolimanesh *et al.*, 2017; Lai *et al.*, 2017; Gao and Wu, 2017). Rural tourism could be this form of tourism suitable for Kalavrita's diversification. As a mountainous area, Kalavrita has both natural and cultural resources as well a strong rural fabric that could be utilized for the suitable development of rural tourism products. As Sharpley (2002) argues, the formulation and implementation of rural tourism could add several benefits and offer important solutions to many rural areas (i.e. creation of new jobs directly and indirectly connected to rural activities, repopulation of rural areas, environmental enhancement, and so on). Indeed, rural tourism can be the key for battling seasonality, generating whole-year destinations, growth in rural areas, tighter horizontal and vertical linkages, promotion of local produce, and higher multiplier effect (Egbali and Nosrat, 2011; Hernández *et al.*, 2016; Boukas, 2019). In the same way, rural tourism necessitates the incorporation of the local and family entrepreneurship, supporting in this way locality and the local society, environment and well-being (Gao and Wu, 2017; Boukas, 2019). Finally, the incorporation of the natural environment is part of the rural tourism nature and, in contrast to mass forms of tourism, creates less stress on natural resources (Egbali and Nosrat, 2011). The multiple contribution of rural tourism is therefore obvious in many dimensions; for this reason, it is suggested in this chapter that it be incorporated in the product/service mix of Kalavrita, as well as other western European destinations with similar characteristics.

Fig. 7.1. Municipality of Kalavrita.

7.3 Kalavrita as a Tourist Destination

Kalavrita is a Greek mountainous town located in the north-west part of Kalavrita area in Peloponnesus, built in Chelmos mountain (735 metres above sea level), approximately two hours from Athens (Fig. 7.1).

It belongs to Achaia Prefecture, western Greece, and it is administered by the Municipality of Kalavrita (Municipality of Kalavrita, 2020). In spite of its natural diversity (mountains, forests, rivers and one of the largest ski resorts in Greece), cultural landscape (various monuments, historic monasteries and archaeological treasures), and historic past (one of the most dramatic towns in terms of loss of life during World War II), the region has only been partially developed. Characteristically, and despite the lack of accurate statistics for the Kalavrita area, the whole western Greece region received only 2.1% of the total tourist overnights in 2018 (approximately 1,915,540 of the overall 91,569,437 overnights) (Hellenic Statistical Authority, 2019). For the Achaia Prefecture, specifically, 30,123 international tourists visited the area in 2018, completing 227,012 overnights and spending 15,653 million euros (INSETE, 2020). Considering that in 2018 almost 28.7 million international tourists visited Greece (Luty, 2019), the percentage that is occupied by Achaia Prefecture (and presumably the Kalavrita area) is significantly small.

These figures signify that even though traces of tourism development exist in the area, in comparison to other Greek tourist destinations such as the Greek islands, the mainland tourist destinations and especially the mountainous ones, such as those in Peloponnesus, it is not developed in the same manner (in terms of development and growth but also in attention from national/regional/local tourism organizations), and could have potential to develop further and become competitive based on appropriate tourist models. Kalavrita is a representative area where even though natural and cultural resources exist as well as the willingness of the local authorities and stakeholders to boost tourism product development, maturity of its tourism product/service mix is not still achieved.

7.4 Methodological Approach

To answer the aims and objectives of the chapter, the study utilized a qualitative research approach. Qualitative research was judged to be an appropriate research method for giving an in-depth analysis of the issues under consideration (Yin, 2009). Data collection incorporated five semi-structured in-depth interviews with local policymakers (representing the overall local policy as well as tourism in the region), community officials, as well as officials on behalf of the tourism sector on the area, during August 2020. A purposive sampling method was judged to be an effective way to derive insight from the most accurate sources. Specifically, different actors of the interaction were interviewed for this chapter: an official from the highest level of the local government (Municipality of Kalavrita) (Respondent 1), a high-ranking representative of the local community (Respondent 2), two high-ranking officials representing the tourism sector in Kalavrita (Respondents 3 and 4) and a publisher/journalist of a large local publication in the area (Respondent 5). All these informants have either a crucial position in the decision-making regarding tourism or know first-hand how tourism development in the area is formed and practised.

The interviews took place in the interviewees' offices and lasted, on average, 60–90 minutes. One interviewer conducted and recorded the interviews. An interview guide was utilized to give a general direction to the interviews and bring to the surface important matters and themes. The main areas that the guide covered were related to the current situation of tourism in the mountainous area of Kalavrita, the definition and importance of rural tourism development in the area, and the strategies for the appropriate implementation of specific rural tourism forms in destinations with the characteristics of Kalavrita. Finally, the interview transcripts were transcribed verbatim and then translated into English from Greek, and the researcher examined the transcripts thoroughly along with documentation written about the destination and underlined important sections to be considered in the analysis of the results. In the following sections, findings that emerged from the thematic analysis and which shed light on the developmental processes and characteristics of Kalavrita mountain region as a tourism destination are presented. The findings are then discussed in relation to the future of rural tourism and the potential to foster the alternative sustainable regional development of tourism in Kalavrita region as well as generally on mountainous destinations with similar characteristics.

7.5 Rural Tourism in Kalavrita: Analysing the Situation

7.5.1 The tourism product of Kalavrita and identified markets

The tourism product of the Kalavrita area is based on its natural and cultural resources as well as older and newer history. Accordingly, its tourism management and marketing authorities capitalize on the region's heritage as well as natural morphology, providing a range of special-interest or alternative tourist activities that take place mainly during the winter months. More specifically, the area offers tourism products that focus on winter tourism and recreation activities (i.e. skiing, rafting, walking on natural trails), cultural heritage, and religious and historic tourism, while the town of Kalavrita possesses one of the oldest rack railway services called 'Odontotos' (established in 1896), which covers a route of 22 km between the town of Kalavrita and Diakopto village and takes a picturesque route through the area (GNTO, 2020). The train brings travellers to the town where they can find cultural, gastronomic and traditional products. As Respondent 5 argues:

> 'Kalavrita is mainly a winter destination since it offers one of the largest ski centres in Greece where tourists can enjoy different winter sports and activities. The entire area is offered for nature-based tourism that takes place on the green paths of the place as well as the Vouraikos river.'

Similarly, another respondent (Respondent 3) mentioned the significant cultural heritage and historic past of the entire area:

> 'There are the historic monasteries of Agia Lavra and Megalo Spilaio where many people travel mostly for religious and historic purposes. Additionally,

> Kalavrita is a place of reference of the Greek history due to the Nazi Atrocity [where 1101 people were murdered by the Nazis in 1943]. Many people visit the execution place as well as the Municipal Museum of the Kalavritan Holocaust to honour the victims of the Nazis.'

In this respect, the tourism product/service mix encompasses experiences related both to the character of the area as well as the cultural mosaic of it.

Regarding the markets that visit the area, respondents indicated that there are various travellers in Kalavrita, but the main market appears to be domestic tourists that visit the place for the weekend. As Respondent 4 claimed:

> 'People that visit the place belong to all age and social groups...the younger people are mostly attracted by the ski centre, the natural and sport activities, while the older are mostly attracted by the religious and historic elements of the area. In any case, there is no clear or absolute distinction since we often see that people would like to meet other types of tourism...Kalavrita can offer diverse tourism forms.'

On the same lines, Respondent 4 suggests:

> 'The most important markets are those people that visit the place for one or two nights, mostly during holidays and long weekends. These are mostly people between 25 and 65 years old, couples, families and groups of friends, usually guests of the ski centre. Groups organized by travel agencies are also visiting and they usually stay one day...'

Respondent 3 states:

> 'We can say that we have three tourist periods: the high period from December till March that we have the most arrivals and overnights mainly from Greek visitors to the ski centre, the low period from April till the end of September where the occupancy rate for the hotels that remain open falls under 40% and mostly represented by international tourists or organized groups, and the middle period from October till end of November where Greeks travel during the weekends and the occupancy rate is better, approximately 60% to 70%.'

Findings indicate that in contrast to mass destinations of Greece that are dominated by international tourists during the summer, Kalavrita is dominated by Greek visitors and tourists that travel individually to the area with their own private car (instead of the train or bus) and visiting during the winter. This indicates reverse signs of seasonality (winter instead of summer) that are encountered in Greece. Overall, all respondents agreed that international tourists could be a more desirable clientele for Kalavrita's tourism – mainly due to their foreign exchange – but their numbers are very low in comparison to domestic travellers. As stated (Respondent 4):

> 'For 2017, tourist arrivals were approximately 27,000 people. From them 24,500 were from the Greek market and only 2500 were internationals ... the greatest percentage of international tourists come from the United Kingdom, the Netherlands, Germany, France and Italy. Of course, we have also the presence of expatriated people from Australia and USA.'

Furthermore, regarding tourist segments, another respondent (Respondent 2) representing the local community appears to be sceptical:

> 'Unfortunately, most of the visitors are "tourists of the one day" ... they go for a walk to the ski centre, then they visit a good "taverna" [Greek type of restaurant], while it is unknown how many of these people visit the Museum of Holocaust...'

The respondent concludes that while during the past many of the tourists used to be religious tourists visiting the historic monasteries of the place, today this model of tourism is seriously weakened. As such, based on the findings, many products in the area, even if they exist, are not fully utilized, appreciated or even explored by the existing tourism markets visiting the place.

7.5.2 Tourism contribution and linkages

Tourism is considered by all respondents as a vital economic activity for Kalavrita town as well as the entire region, especially during the last decades. As mentioned, 'Tourism in the area of Kalavrita municipality is a beneficial activity for the local community' (Respondent 4). This happened mainly due to the economic shift from other traditional sectors of the economy to tourism. As Respondent 2 argued:

> 'In older times, Kalavrita was full of people and an intensive economic activity mainly around farming and livestock was noted. Then, urbanization happened and many of these activities were slowly abandoned ... people, especially our younger ones migrated to other urban areas of Greece, such as Athens, to find jobs.'

The same respondent argues that a solution for this problem was the transformation of the local economic activity to tourism, utilizing mostly the mountainous regional landscape.

Indeed, tourism today plays a central role in the overall growth of the place. As Respondent 1 suggested:

> 'All the activities [tourism activities] are tightly connected to the general area of Kalavrita, cover the natural elements of the place, and for this reason are projected and promoted [by the tourism policy] in order for the visitation and recognition to be achieved, aiming [at] the economic and overall development of the area.'

On the same lines:

> 'During the recent years, tourism has played a very important role for the area and has helped its economic development. For instance, it has created many jobs for local people, it has led to the improvement and construction of infrastructure offering a good quality of life to tourists as well as local residents, it has helped the global and cultural relationships of different people, broadening this way the spiritual horizon and the residential development.' (Respondent 5)

According to the interviews, tourism's significance can be indicated by its contribution to other industries such as farming and agriculture, leisure and trade. As Respondent 2 mentioned:

> 'As many of our people have moved to the cities to find jobs, many of the traditional activities have been abandoned ... of course, tourism gave an extra income to the area and helped also the other economic activities to grow through the promotion, for instance, of our local produce such as the Kalavrita's feta [cheese] or through the attraction of customers to restaurants and shops ... without tourism many of these companies would be closed.'

The importance of other industries is also signified by tourism officials of the area as many of the stakeholders are important producers of the tourism product/experience of Kalavrita and therefore they (tourism officials) try to enhance their relationships. As stated by Respondent 3:

> 'Tourism as an amalgam of many and diverse services and tangible products enacts dynamically to all the sectors of production and for this reason, the co-operation and interaction with local government, local stakeholders, cultural associations, mountaineering clubs, the Management Body of Chelmos Vouraikos, the presidents of local communities, the commercial association as well as the hotel association and the Prefecture of Western Greece, is a necessary precondition for the development of the tourism product of the entire area, since all have as a common aim the development and enhancement of the place.'

Apart from the economic contribution of tourism, respondents indicated also its social contribution by signifying the history of the place to travellers as well as the better appreciation of the natural environment through the engagement of tourists with natural-based tourism actions and events. It appears, therefore, that as with many other destinations, tourism stakeholders and policy valorize the significance of tourism and many of them have a direct contribution to the tourism experience offered in the surrounding area.

7.5.3 Strengths and weaknesses of Kalavrita

Based on the findings, Kalavrita could be characterized as a destination with many qualities that can offer multiple and varied experiences to tourists and visitors. Undoubtedly, the major strength of Kalavrita area is its natural environment and its multi-dimensional cultural resources. This is something highlighted by every respondent. Indeed, a plethora of natural trails, unique environment and morphology as well as valuable cultural heritage wealth have played a vital role in the development of tourism.

Another advantage underlined by the interviewees is the fact that Kalavrita is located in close proximity to the capital city of Greece, Athens (approximately 200 km), and Patra (73 km), the third largest city in Greece. As Respondent 4 argues:

> 'Kalavrita's tourist competitiveness has been improved … apart [from] being in close proximity to great cities of Greece, a very crucial role towards this direction has been played by the completion of the highway that connects Athens and Patra for easy accessibility to the area [Kalavrita is located close to Patra]. The feeling of safety in road connections is a strong incentive for choosing destinations, especially for international tourists. Moreover, the quick connection with the first airport of the country, Eleftherios Venizelos Airport, and the first sea port of Greece in Patra, that connects the country to the rest of Europe, are also of great importance.'

It appears, therefore, that both the resource deposits of the area as well as its location are strong points for Kalavrita's product/service mix. Another respondent (Respondent 5) mentioned that there are also many opportunities for further enhancement of Kalavrita's offering. For instance, the respondent indicated the example of the Chelmos Planetarium that could attract visitors with an interest in space and allow them to participate in relevant activities. Similarly, Respondent 1 mentioned the rack railway service, which could be an attraction on its own, or Respondent 2 highlighted the rich primary sector of the area that could be a basic source of experiences for farming and agricultural tourism development in the area. Nonetheless, as stressed, and will be further discussed, many of these opportunities are not fully seized.

Regarding the weaknesses, some of the interviewees stressed the ephemeral character of the existing tourism product which is mainly planned and developed based only on winter tourism (especially with emphasis on the activities that take place at the ski centre). As stated, much of the activity in the area mainly targets Greek visitors and tourists that would go for a weekend to the place, during the winter, mainly for skiing. As Respondent 5 argued:

> 'Tourism development has transformed Kalavrita into a one-season destination … there is a plethora of accommodation units most of them remain closed or they are malfunctioning while there is no plan in any type of development.'

Indeed, seasonality is also existent in Kalavrita where, as stated, many hotels are full during the weekends but empty during the week, while many of them are not operating at all during the non-peak period (summer months). A direct outcome of this situation is the unequal development of tourism where certain activities in specific areas (i.e. skiing) receive the greatest percentage of tourists and thus the attention of the local management/marketing authorities, whereas others (i.e. cultural, religious, historic) are marginalized or exist and operate subtly only as secondary products/services peripheral to the main tourism product/service mix, rather than as autonomous tourism offerings with their own character and marketing management. In this respect, the benefits of tourism are not equally spread to the entire area and the various economic sectors. This unequal distribution creates unbalanced development where some benefit much more than the majority of the population, even though the last one is also a significant producer of the 'mountainous tourism experience'. As mentioned: 'We live [in terms of tourism] only because of the ski centre' (Respondent 2).

On top of that, despite improved accessibility due to the upgraded highway that connects Athens with Patra, the regional road from the highway to Kalavrita is still problematic. As Respondent 4 adds: 'We are not very proud of the provincial roadway ... there are many problems noted that can make the route dangerous, especially during the winter.' Indeed, accessibility appears to be an important problem for many mountainous destinations with similar characteristics to this case study. Easy accessibility is important for the holistic tourism product development of the area as it does not only provide easier access to the mountainous area but also safety while travelling to and from the destination.

Generally, according to respondents, weaknesses of the area's tourism have to do with: its nature (mainly seasonal, and thus tourism is unequally spread with low and peak seasons); its quick development mostly for short-term profitability and frequently unplanned, generating overall a weaker tourist experience; the creation of a situation where many locals have shifted away from traditions and customs; the burdening of the natural environment of the area; employees' insecurity due to seasonality; the cutting down of forests for the development of tourist facilities; the decline of farming and other activities; and the weakness in local interrelationships (Respondents 3 and 5). All these weaknesses harm seriously the profile, character and identity of the destination's product/service mix and generate questions regarding the quality of the obtained tourist experience.

Furthermore, respondents mentioned also the attitude of stakeholders towards tourism that even though there is potential for further and diversified tourism development, important opportunities are being lost. As Respondent 1 argues: 'Our greatest enemy is ourself and our tendency to want easy and quick money. We want to become rich without even trying.' Similarly, Respondent 2 claims: '[M]any times local governance does not pay any attention at all' [in regards to tourism development]. Finally, Respondent 5 argues that:

> 'We could do [much] more with our history, religion or even the traditional railway service ... in contrast, religious tourists are turned away because we don't promote as we should our historic past and the railway is mostly seen as a means of transportation rather than an attraction that can offer in a limited time unique and different experience in the nature of Kalavrita.'

Overall, based on the interviews, Kalavrita, even though characterized by a series of features that could play a significant role in the creation of a balanced competitive tourism product, is mainly developed around one theme (winter and ski tourism) that systematically excludes product attributes that could offer added value, diversification, enrichment and deeper experiences, capitalizing on its rurality. In this respect, rural tourism was discussed by the interviewees.

7.6 Rural Tourism in Kalavrita: Challenges, Opportunities and the Need for Diversification

According to respondents, there are several forms of rural tourism that take place in the Kalavrita region, such as winter tourism, sport and recreation tourism, cultural heritage tourism, religious tourism, etc. As noted from the interviews, even though these micro-rural activities are developed and operate in the area, there is no systematic and serious rural tourism planning, development and management. One of the reasons for this situation is the complexity of rurality itself: 'It is very difficult to discuss organized rural tourism development in an area where everything is rural' (Respondent 4), but probably the most discussed reason is because certain tourist activities (i.e. ski tourism) dominate over many others (and hence receive the greatest attention both by tourism authorities and professionals). Therefore, balancing an existing situation appears to be difficult for the area's tourism stakeholders. In marketing terms, this also raises issues regarding branding and identity of the area's tourism as many times there is no consistency in regard to tourist offering, since several activities are offered only in a limited way in comparison to others (e.g. cultural tourism over ski tourism).

Additionally, the requirement for many and different people that need to participate in the production and formulation of the rural tourism experience appears to be also challenging. This is even more imperative since most of the forms met in the area represent individual efforts and projects. As mentioned by Respondent 3, the co-operation among different stakeholders needed for the proper implementation of rural tourism activities is an encounter that many times turns the initiative away:

> 'At the beginning of any attempt we meet difficulties and problems that concern mainly the collaboration, sympathy and readiness of all the stakeholders for the achievement of the goal but also for the maintenance and improvement of the tourism product in order for it to be competitive.'

And similarly:

> 'A necessary prerequisite for transforming Kalavrita into a whole-year destination is the co-operation of all stakeholders such as the regional authorities, municipality, professionals, etc., and the creation of an organized strategic business plan for the development and promotion of Kalavrita from tourism professionals which will be the guide for any additional actions.' (Respondent 4)

The difficulty for the full co-operation of tourism stakeholders in the area in combination with the unequal spread of benefits around it, is also an important reason that hinders rural tourism development:

> 'On the one hand, locals do not have the necessary knowledge or even communication [regarding tourism development] that is necessary for rural tourism, and, on the other hand, they see that most of the money goes to tourism around the ski centre. Obviously, this is not an incentive for them to deal seriously with tourism.' (Respondent 2)

In the same manner, respondents mentioned that the benefits of tourism should be spread equally, not only to the town of Kalavrita but also to the smaller villages where nature and resources are still in their infancy and could be sustainably developed (Respondents 2 and 5). Based on these findings, it appears that one of the main producers of the rural tourism experience, the local community in the surrounding area of Kalavrita, has a weakened role in regards to tourism product development, and thus its contribution is not fully recognized and/or appreciated.

Additionally, the contemporary landscape in tourism around the world generally, and Greece specifically, does not help the situation:

> 'Tourism, despite the economic crisis that hit the country during recent years and the intense competition, needs to be constantly in vigilance and continuous improvement and renewal in the services that it provides.' (Respondent 5)

This is even more essential considering that modern tourism will soon enter the post-Covid-19 era.

In line with the above, the fragmented nature of the rural tourism product seems to be also a noteworthy condition that confuses the stakeholders: 'Even if we talk about rural tourism, there are many forms of it that happen in the area but also many others that do not ... this creates lost opportunities' (Respondent 3). The respondents mentioned that there is still much space for further development for many other new rural forms of tourism (for the area), such as agricultural tourism (on an organized basis and in close co-operation with the primary economic sector), gastronomic tourism (incorporating both preparation and consumption), health and healing tourism, adventurous tourism, etc. (Respondents 3 and 4). Some more specific ideas mentioned in the interviews included: the creation of sport tourism (other than on snow); the further enrichment of culinary tourism with local components such as the traditional feta cheese or the trout of the Planeteros area; the creation of 'walking tourism' within certified nature trails, as this attracts many international tourists; the creation of events such as cycling events with mountain bikes and downhill motorcycle races; cultural events; and organized activities in the rural region. Of course, the interviewees agree that in order for this to happen, there is a need for further study, co-ordination and communication among different service providers.

Furthermore, respondents also mentioned innovation towards this goal: 'Since the majority of tourism business that happens in the area comes from small and medium-sized entrepreneurs, we need to find innovative ways and processes to deal with our current and tourism forms', and 'We need to adapt to the new needs and challenges of today's era by capitalizing on the digital technology and upgrading our tourist information' (Respondents 4 and 3, respectively). Towards this goal, tourist education and consciousness enhancement are a necessity for achieving this: 'We need to motivate the local community for the development and provision of qualitative tourist services, through the training of all employees in tourism and at all levels for the creation of a skilled workforce with up-to-date competences' (Respondent 4).

This is something imperative as product quality was an issue mentioned as a weakness regarding the regional tourism of Kalavrita.

Finally, the diversification and enrichment of the tourism product involves the maintenance, improvement and development of the infrastructure and superstructure that would improve the quality of life of locals but always in harmony with the support of the tourism product (Respondent 2). Overall, respondents mentioned that for the appropriate adaptation of an organized model of rural tourism development there is the need for an office of tourism that would have a central role in the systematic promotion and projection of the tourism product of the entire area of Kalavrita. This office should be responsible not only for the creation and co-ordination of existing and new tourism products in the region but also for the modern destination marketing through digital tools that would offer constant and pertinent information not only to domestic but also, more necessarily, to international tourists.

7.7 Discussion, Implications and Conclusion

According to the findings, the mountainous area of Kalavrita is a destination where many and diverse rural tourism forms can grow and develop since it is characterized by all the conditions for rural tourism development (Lane, 1994). In this respect, the unique natural and cultural heritage resources and the balanced mix of attractions and facilities, spread throughout the area (Kastarlak and Barber, 2013), as well as the local character of the business environment in the area, are important determinants for its further success. Yet there is a number of challenges that still prohibit this further growth. As seen in the study, these include seasonality, lack of and complicated synergies among stakeholders (that have a crucial role in the production of the rural tourism experience), monoculture of certain tourist activities and underestimation of others, inefficient tourism planning, and infrastructure problems.

As derived from the findings, these chronic problems, while evident, have not been handled by tourism management and marketing authorities, and thus are still remaining today. In this respect, while Kalavrita could develop a diversified tourism product, based on its natural and cultural resources (tangible and intangible), that could operate during the whole year and in several places, its focus is more mono-dimensional towards winter tourism and without utilizing available activities and resources. This stresses the importance for the product diversification, the enrichment of the tourism product portfolio of the entire area, and the sustainable development and management of tourism benefiting the local economy, society and, obviously, environment. In this respect, rural tourism can offer to the entire area of Kalavrita the opportunity to become sustainably competitive, as most of its forms seen in the area are, by nature, sustainable (or more sustainable in comparison with the conventional tourism product in Greece).

Initially, then, existing challenges need to be faced in order to deal systematically and in an organized way with rural tourism. Some of these are local to the tourism of the area, but others apply to many other rural mountainous destinations. For instance, accessibility, as mentioned in this case study, is characterized as one of the main problems in tourism development of the mountainous areas of Cyprus (Boukas, 2019), of Nafpaktia in Greece (Bitsani and Kavoura, 2012), or in the mountainous areas in Cuba (Apollo and Rettinger, 2019). Placing emphasis, therefore, on the improvement of the infrastructure/superstructure, especially for mountainous, difficult-to-visit destinations, is imperative in order for more tourists to be able to easily visit them and live the rural tourism experience.

Additionally, in order for the diversification and enrichment of the current tourism offering to be efficiently managed, there is a need for the formulation of a strategic tourist plan that would incorporate goals and strategies that, on the one hand, would lead to the competitiveness of the destination but also, on the other hand, to its sustainability, spreading the benefits from tourism in different directions: economy, society, environment (Kastarlak and Barber, 2013). It appears that the lack of a strategic plan is one of the main problems noted in the study. Indeed, according to Pettas (2013), the area of Kalavrita has been mainly developed in terms of tourism during the last decades, but this development was not part of holistic tourism planning. As such, the author agrees with the findings and argues that the spatial distribution of the tourism activities is unequal and without any strong backward linkages. Hence, while Kalavrita is a famous winter destination during the weekends (Sgartsou, 2007), any development that takes place is mostly based on a specific profile of tourists (ski lovers) for a specific period of the year (winter). Pettas (2013) concludes that despite the competitive advantages of the destination, in terms of physiognomy and cultural/natural heritage, the lack of a complete plan that could incorporate other sectors of production (mainly primary sector) and lead to more sustainable development is obvious, and stresses the need for change of the current situation.

Towards this end, a series of suggestions are proposed in this chapter. A common theme derived from the findings is the collaboration among different stakeholders. As mentioned, the rural tourism product is very fragmented and produced by different providers from different industries and from both the public and private sector. In this regard, therefore, an appropriate strategy is the co-ordination and co-operation of these people for the formulation of a proper rural tourism experience. Synergies among rural tourism stakeholders is indeed a significant prerequisite for the successful rural tourism of an area. For instance, as Van Huylenbroeck *et al.* (2006) argue, for the case of agricultural tourism, the attractiveness of a rural destination is enhanced by the contribution of farming while farmers also benefit from tourism by offering accommodation or selling their merchandise to tourists. The authors argue that building strong relationships among tourism and farming is an efficient strategy both for countryside management and rural development.

For these synergies to be achieved, however, there is a need for trust and mutual spread of benefits. As Quaranta *et al.* (2016) argue, many times rural tourism development is not an easy task to be performed: various rural businesses are remote, the local market is characterized by peripheralization in contrast to the urban centres where the majority of the tourism demand is located, while the highly fragmented and difficult to co-ordinate rural activity is evident. These problems create an environment where many people in the rural areas feel disconnected to the rest of the world and this results in lack of confidence and belief in how they view themselves. To tackle this situation the authors claim that it is crucial to recognize shared spaces where rural actors can negotiate and decide about their vision regarding community development. According to them, it is important to rebuild 'community ties and promote a greater sense of community through a long-term, inclusive, transparent, and democratic approach' (Quaranta *et al.*, 2016, p. 12). This can be achieved by eliminating cultural barriers and stimulating the appropriate opportunities for synergies among various sectors. Participation in rural tourism development could be one of these opportunities for synergy. A consequence of a strategy like this is the positive attitude of locals to be more passionate and active towards the formation of a qualitative rural tourism product, offering diversified tourist activities and genuine experiences to visitors. By this strategy, locals become main actors to the tourism offering, gain a better understanding by tourism marketing about their contribution to product development, and hence their role is better appreciated both by tourism authorities and tourists themselves as receivers of the rural experience.

Additionally, education of the local community and stakeholders was mentioned as a necessity in order for the conditions of rural tourism development to exist. For rural tourism to work, there is a need for the identification of stakeholders and their education for the proper and equal participation in decision-making and collaboration (McComb *et al.*, 2017). The suitable synergies and the structure of a common environment where rural tourism can grow and spread fairly is a necessity. In order, therefore, for this to happen there is a need for an organized rural tourism body (tourist office) that would take into consideration all the relevant stakeholders/producers of the experience, would constantly train and develop them through the help of tourist professionals, and, ultimately, would incorporate them in the area's sustainable tourism marketing planning. Then, a rural tourism portfolio of the area could be developed where any rural tourism activity in all the areas of Kalavrita (both the town as well as the villages) would be included, while guidance and consultation to each business that needs it could be provided.

Hence, by emphasizing rural tourism, more appreciation to the local fabric (in terms of destination attributes as well as community enhancement) can lead to a better involvement of the local element in the production process. To this end, tourism marketing needs to take initiatives for strengthening and highlighting the rural tourist experience. For instance, this could be achieved by adopting innovative communication techniques such as interpretative centres spread over the area (Carter and Grimwade, 1997), or creating more functional rural products that would give the opportunity to visitors to interact better with the destination's features (e.g. agricultural products production, cultural traits, natural heritage elements, etc.), signifying the uniqueness of the place, telling a story, or trying to engage the traditional character of the past to the modernized foundations of the present (i.e. prompting memories, demonstrations, active participation, etc.) (Misiura, 2006; Boukas, 2012).

Similarly, tourism marketing ought to build a stronger long-lasting rural tourism brand that would communicate a solid identity of the area (Evans, 2003), not only as a ski destination but also as a region where someone can experience diverse qualitative rural tourism experiences, in a unique natural and cultural heritage landscape. The establishment of a sturdy brand needs to be part of a holistic marketing plan that would embrace the destination's specific quality and create a consistent image or set of images in the minds of the visitors/tourists (Boukas, 2013). These images should be formed in combination with a series of the destination's tourism attributes, paying close attention to the local element of the area, to its heritage, and the traditional character of the region's life. This branding would give value on its own to the destination, through brand equity. As Misiura (2006) argues, a brand's asset value/equity is important for heritage destinations. Considering that Kalavrita is a destination with significant heritage, this step appears to be an imperative marketing strategy that could tackle many challenges in the area.

In order for this rural product and identity to be established it is imperative to consider also the perceptions of visitors and tourists that visit Kalavrita. In this respect, the proposed tourist authority would be responsible for marketing of rural tourism in the area to conduct systematic tourist behaviour research that would include several dimensions such as tourist perceptions, motivation, attitudes and satisfaction with the rural tourism element of the area. Tourists' perceptions are a significant aspect that can influence the tourists' destination choice (Wong Chak Keung, 2000) since the elements that tourists perceive as important for a destination can influence them positively or negatively towards that destination. Similarly, satisfaction surveys are a strong marketing tool that can link processes like purchase and consumption with post-purchase phenomena such as attitude change, repeat purchase, and brand loyalty (Churchill and Surprenant, 1982). Besides, overall satisfaction can predict efficiently customer behaviour and intentions (Olsen and Johnson, 2003), therefore its significance is great. In this respect, using diverse methodologies on understanding the views of tourists regarding the rural tourist offering of the area is imperative for the future marketing of the entire region.

To conclude, rural tourism in mountainous areas can exist only when all members of the community see and enjoy its virtues but also share its costs. This requires the formulation of a long-term marketing vision and the active participation of all stakeholders/key players in the production, distribution and promotion of the rural tourism experience. Product diversification, in this case, does not include only the new products developed, but also innovative processes (e.g. using new technologies in the production process, co-creation of experiences, creative tourism, etc.) that would help these products to grow and be fertile, offering equally spread benefits to many and diverse stakeholders. Further research should investigate these processes and test them to measure their effectiveness. Another field for further study is to explore how the transformation of the contemporary tourism landscape that is affected by a series of global forces (i.e. health crises, economic recessions, globalization, etc.) can lead to the more efficient marketing of rural tourism destinations such as Kalavrita.

Among the limitations of the study is the sampling used, interviewing only five people representing mainly the supply/policy/marketing side of the tourism provided in the area. Studying also the views of tourists that visit the destination regarding their experience from a series of destination attributes would provide insight to tourist marketers for the more efficient management of the offered product in the destination. Similarly, findings would be enriched if the qualitative study incorporated various rural tourism providers in the area.

Overall, the study indicated that many rural areas, despite the demand trends for 'pure experiences' and the amounts of resources in the rural environment, lack proper management (including marketing) and hence fail to be competitive as well as sustainable. Considering that sustainable development is the future of tourism and that rural areas can offer these characteristics to achieve it, developing and maintaining a portfolio of rural tourism forms could be a proper marketing solution for the long-term viable and sustainable tourism product development of many mountainous areas such as Kalavrita.

Acknowledgement

My sincere gratitude is expressed to Major General Sotirios Tsenes for his valuable contribution to this chapter.

References

Apollo, M. and Rettinger, R. (2019) Mountaineering in Cuba: improvement of true accessibility as an opportunity for regional development of communities outside the tourism enclaves. *Current Issues in Tourism* 22(15), 1797–1804. DOI: 10.1080/13683500.2018.1446920.

Benur, A.M. and Bramwell, B. (2015) Tourism product development and product diversification in destinations. *Tourism Management* 50(1), 213–224. DOI: 10.1016/j.tourman.2015.02.005.

Bitsani, E. and Kavoura, A. (2012) Accessibility versus advertising for mountain tourism: the case of Nafpaktia. *Tourismos* 7, 217–236.

Boukas, N. (2012) Young faces in old places: perceptions of young cultural visitors for the archaeological site of Delphi. *Journal of Cultural Heritage Management and Sustainable Development* 2, 164–189. DOI: 10.1108/20441261211273653.

Boukas, N. (2013) Youth visitors' satisfaction in Greek cultural heritage destinations: the case of Delphi. *Tourism Planning & Development* 10(3), 285–306. DOI: 10.1080/21568316.2012.747986.

Boukas, N. (2019) Rural tourism and residents' well-being in Cyprus: towards a conceptualised framework of the appreciation of rural tourism for islands' sustainable development and competitiveness. *International Journal of Tourism Anthropology* 7(1), 60–86. DOI: 10.1504/IJTA.2019.098105.

Carter, B. and Grimwade, G. (1997) Balancing use and preservation in cultural heritage management. *International Journal of Heritage Studies* 3(1), 45–53. DOI: 10.1080/13527259708722186.

Churchill, G.A. Jr. and Surprenant, C. (1982) An investigation into the determinants of customer satisfaction. *Journal of Marketing Research* 19(4), 491–504. DOI: 10.1177/002224378201900410.

Daspher, K. (2014) *Rural Tourism: An International Perspective*. Cambridge Scholars Publishing, Newcastle upon Tyne, UK.

Egbali, N. and Nosrat, A.B. (2011) Effects of positive and negative rural tourism (case study: Rural Semnan Province). *Journal of Geography and Regional Planning* 4, 63–76.

Evans, G. (2003) Hard-branding the cultural city: from Prado to Prada. *International Journal of Urban and Regional Research* 27(2), 417–440. DOI: 10.1111/1468-2427.00455.

Gao, J. and Wu, B. (2017) Revitalizing traditional villages through rural tourism: a case study of Yuanjia Village, Shaanxi Province, China. *Tourism Management* 63(144), 223–233. DOI: 10.1016/j.tourman.2017.04.003.

GNTO (2020) Kalavryta. Available at: http://www.visitgreece.gr/el/mainland/kalavryta (accessed 10 September 2020).

Hall, D., Roberts, L. and Mitchell, M. (2003) *New Directions in Rural Tourism*. Routledge, Abingdon, UK.

Hellenic Statistical Authority (2019) Arrivals and nights spent in hotels, similar establishments and tourist campsites: 2018. Press release. Trade and Services Statistics Division, Tourism Statistics Section.

Hernández, J.M., Suárez-Vega, R. and Santana-Jiménez, Y. (2016) The inter-relationship between rural and mass tourism: the case of Catalonia, Spain. *Tourism Management* 54(2), 43–57. DOI: 10.1016/j.tourman.2015.10.015.

Hjalager, A.-M. (1996) Agricultural diversification into tourism: evidence of a European Community development programme. *Tourism Management* 17, 103–111.

Hjalager, A.-M., Kwiatkowski, G. and Østervig Larsen, M. (2018) Innovation gaps in Scandinavian rural tourism. *Scandinavian Journal of Hospitality and Tourism* 18(1), 1–17. DOI: 10.1080/15022250.2017.1287002.

INSETE (2020) *Key Figures of Incoming Tourists in Western Greece*. INSETE Intelligence, Athens.

Kastarlak, B.I. and Barber, B. (2013) *Fundamentals of Planning and Developing Tourism*. Pearson, Hoboken, New Jersey.

Lai, P.-H., Morrison-Saunders, A. and Grimstad, S. (2017) Operating small tourism firms in rural destinations: a social representations approach to examining how small tourism firms cope with non-tourism induced changes. *Tourism Management* 58(2), 164–174. DOI: 10.1016/j.tourman.2016.10.017.

Lane, B. (1994) What is rural tourism? *Journal of Sustainable Tourism* 2(1-2), 7–21. DOI: 10.1080/09669589409510680.

Li, S.X. and Greenwood, R. (2004) The effect of within-industry diversification on firm performance: synergy creation, multi-market contact and market structuration. *Strategic Management Journal* 25(12), 1131–1153. DOI: 10.1002/smj.418.

Luty, J. (2019) Number of arrivals in tourist accommodation Greece 2006–2018. Available at: https://www.statista.com/statistics/413222/number-of-arrivals-spent-in-short-stay-accommodation-in-greece/ (accessed 10 September 2020).

McComb, E.J., Boyd, S. and Boluk, K. (2017) Stakeholder collaboration: a means to the success of rural tourism destinations? A critical evaluation of the existence of stakeholder collaboration within the Mournes, Northern Ireland. *Tourism and Hospitality Research* 17(3), 286–297. DOI: 10.1177/1467358415583738.

Misiura, S. (2006) *Heritage Marketing*. Butterworth-Heinemann, Oxford.

Municipality of Kalavrita (2020) Municipality of Kalavrita. Available at: http://www.kalavrita.gr (accessed 10 September 2020).

Olsen, L.L. and Johnson, M.D. (2003) Service equity, satisfaction, and loyalty: from transaction-specific to cumulative evaluation. *Journal of Service Research* 5(3), 184–195. DOI: 10.1177/1094670502238914.

Pappas, N. and Papatheodorou, A. (2017) Tourism and the refugee crisis in Greece: perceptions and decision-making of accommodation providers. *Tourism Management* 63(1), 31–41. DOI: 10.1016/j.tourman.2017.06.005.

Pettas, D. (2013) Tourism development of Kalavrita Municipality: current situation, challenges, and potential. *Kalavrita News* 25 April. [in Greek].

Quaranta, G., Citro, E. and Salvia, R. (2016) Economic and social sustainable synergies to promote innovations in rural tourism and local development. *Sustainability* 8(7), 668. DOI: 10.3390/su8070668.

Rasoolimanesh, S.M., Ringle, C.M., Jaafar, M. and Ramayah, T. (2017) Urban vs. rural destinations: residents' perceptions, community participation and support for tourism development. *Tourism Management* 60(1), 147–158. DOI: 10.1016/j.tourman.2016.11.019.

Roberts, L. and Hall, D. (2001) *Rural Tourism and Recreation: Principles to Practice*. CAB International, Wallingford, UK.

Robinson, M. and Novelli, M. (2005) Niche tourism: an introduction. In: Novelli, M. (ed.) *Niche Tourism: Contemporary Issues, Trends and Cases*. Elsevier, Oxford, pp. 1–14.

Sgartsou, D. (2007) Large tourism development possibilities in Kalavrita area. TravelDailyNews.gr, 9 Octoberin Greek.

Sharpley, R. (2002) Rural tourism and the challenge of tourism diversification: the case of Cyprus. *Tourism Management* 23(3), 233–244. DOI: 10.1016/S0261-5177(01)00078-4.

Stratigea, A. and Katsoni, V. (2015) A strategic policy scenario analysis framework for the sustainable tourist development of peripheral small island areas: the case of Lefkada-Greece Island. *European Journal of Futures Research* 3(5). DOI: 10.1007/s40309-015-0063-z.

Stylidis, D. and Terzidou, M. (2014) Tourism and the economic crisis in Kavala, Greece. *Annals of Tourism Research* 44(3), 210–226. DOI: 10.1016/j.annals.2013.10.004.

Tsilimigkas, G. and Derdemezi, E.-T. (2019) Unregulated built-up area expansion on Santorini Island, Greece. *European Planning Studies* 28(9), 1–22. DOI: 10.1080/09654313.2019.1687656.

Van Huylenbroeck, G., Vanslembrouck, I., Calus, M. and Van de Velde, L. (2006) Synergies between farming and rural tourism: evidence from Flanders. *EuroChoices* 5, 14–21.

Wong Chak Keung, S. (2000) Tourists' perceptions of hotel frontline employees' questionable job-related behaviour. *Tourism Management* 21(2), 121–134. DOI: 10.1016/S0261-5177(99)00042-4.

Yin, R.K. (2009) *Case Study Research: Design and Methods*, 4th edn. Sage, Thousand Oaks, California.

8 US Tourists' Food Neophobia and Their Local Food Experiences in France and Italy

JA YOUNG (JACEY) CHOE*

University of Macao, Macao

8.1 Introduction

Food is not merely a compound of nutrients for providing energy to the human body. Tourists find pleasure, feel excitement, confirm their identity, experience new culture and strengthen family cohesion through food consumption in a destination (Long, 2004; Getz and Robinson, 2014). From the destination marketing perspective, a destination marketing organization puts a lot of effort into enhancing a destination's image through food (Choe and Kim, 2018). Given that tourists like to taste local cuisine belonging to a region, local food is an important way of learning more about a destination's culture (Stanley and Stanley, 2014). However, tourists are hesitant to try local food in a new destination because they are unfamiliar with the taste of novel food and feel uncomfortable with new food culture. Food neophobia, which is one of the most important food-related personality traits, is defined as the reluctance to eat, and/or avoidance of, new food (Pliner and Hobden, 1992).

Given that food neophobia can greatly influence preferences towards novel food (Tuorila *et al.*, 2001), understanding its potential impact on tourists' local food perception experiences in a destination is an important issue for destination food marketers. Nevertheless, topics regarding food neophobia among tourists have been surprisingly under-researched. No study has investigated the associations among food neophobia, tourists' socio-demographics, tourists' local food consumption values and tourists' behavioural intention to return to a destination for its food.

Researchers in the hospitality and tourism field have acknowledged the importance of consumer values (Sánchez *et al.*, 2006; Ryu *et al.*, 2008). However, relatively few studies have investigated consumption value compared with

*jaceychoe@um.edu.mo

DOI: 10.1079/9781789248753.0008

the field of marketing (Petrick and Backman, 2002). Furthermore, the importance and influence of behavioural intention to return to the destination for its food have yet to receive substantial academic attention. In particular, in the context of tourists' local food consumption, few studies have investigated the effect of the food neophobia personality trait on tourists' food consumption values and behavioural intention to return to the destination for food tourism.

In this regard, this study has three major objectives; the first is to explore whether tourists have different levels of food neophobia according to their socio-demographics; the second is to investigate if differences in tourists' local food consumption values exist among different food neophobia groups (FNGs); and the third is to examine whether differences in tourists' behavioural intention to go back to a destination for food tourism exist among different FNGs. This study expects that destination marketers, hospitality businesses and tour operators can achieve a competitive advantage by understanding the concept of food neophobia and its associations with tourists' socio-demographics and local food experiences in a destination. In particular, this study will focus on US tourists' local food perception experiences in France and Italy.

8.2 Literature Review

8.2.1 Food neophobia in tourism

Food neophobia refers to the tendency to reject novel food based on a fear of unfamiliar food (Pliner and Hobden, 1992). Some researchers have argued that food neophobia should be treated as behaviour, involving the avoidance of new food in a particular situation (Pliner and Salvy, 2006). However, most previous studies have indicated that food neophobia is more like a food-related personality trait that cannot be changed easily and lasts for a long time (Nordin *et al.*, 2004; Choe and Cho, 2011).

French sociologist Fischler (1988) introduced the concept of the omnivore's paradox. Humans are omnivores, meaning that various nutrients are required for metabolism and humans can consume a wide range of food. However, humans must differentiate between what is edible and what is not. In other words, eating new foods can be a risky business for humans because some food ingredients could be toxic, causing illness or death (Rozin, 1976). Therefore, from a biological perspective, food neophobia serves a protective function (Zachmann and Østby, 2011). From the perspective of healthy eating, food neophobia has a negative impact on the range of food people eat and the consumption of recommended foods, such as vegetables or fish (Siegrist *et al.*, 2013). Food neophobia is also an important issue for food product developers to overcome when they introduce new food products to the market (Henriques *et al.*, 2009). Thus, understanding food neophobia has been considered important in the fields of food science and consumer research.

In tourism research, a handful of studies have discussed food neophobia (Cohen and Avieli, 2004; Kim *et al.*, 2010; Chang *et al.*, 2011; Mak *et al.*, 2012, 2017; Mkono *et al.*, 2013). However, many of them merely mention food neophobia to support their research framework or conceptually predict its influence on tourist food consumption. Mak *et al.* (2012) applied the concept of the omnivore's paradox suggested by Fischler to the context of tourists' food consumption and explained that the tourist's paradox is an oscillation between pursuing the 'symbolic' and fulfilling the 'obligatory'. The symbolic aspect is related to neophilia (seeking and enjoying novelty), whereas the obligatory aspect is associated with neophobia (fear of novelty). According to Molz (2007), the act of eating new food contains a great level of 'symbolic' value because to do so, consumers, to a certain extent, have to be adventurous, adaptable to a different setting and open to another culture. Similarly, cultural capital theory explains that consuming local food in a new destination is a learning activity through which a tourist can obtain knowledge and experience the local culture. During the last decade, tourists have shown a high tendency of being educated about local food culture to increase their cultural capital (Fields, 2002; Chang *et al.*, 2011). The 'obligatory' aspect of food consumption cannot be ignored in tourism (Mak *et al.*, 2012). Many tourists need an 'environmental bubble' to enjoy new experiences in a destination (Cohen, 1972). Tourists may want to have 'authentic' local food on the condition that they can filter out any aspects of local culinary situations they may find repulsive (Cohen and Avieli, 2004).

Cohen and Avieli (2004) emphasized the importance of researching food neophobia in the tourism context because tourists are greatly exposed to new and unfamiliar food and dishes at a new destination. They stated that tourists can be classified according to the level of their food neophobia. Given that eating local food in a new destination necessitates actual bodily involvement combined with being in an unfamiliar environment, neophobic tendencies are highly likely to become more prominent. Similarly, Torres (2002) mentioned that tourists are comfortable when they eat familiar food in a new destination; thus, a food neophobic tendency may become obvious in a new destination. However, the above studies did not classify tourists into different levels of food neophobia to explore further the impact of food neophobia on tourists' food experiences.

Mak *et al.* (2012) stated that food neophobia is one of the most important psychological variables affecting tourist food consumption. According to their study, food neophobic tourists may show reluctance to try local food in a new destination. In the same way, Mak *et al.* (2017) attempted to examine the effect of food neophobia on tourist food motivations. In their study, food neophobia was found to have significantly explained variance for novelty and variety, interpersonal and cultural, health concern and familiarity motivation factors. Mak *et al.*'s (2017) study is one of the few studies that provided empirical evidence for the impact of this food-related personality trait on tourists' food consumption. However, they did not focus on tourists' actual evaluations or perceptions of local food consumption in a destination.

Moreover, Kim *et al.* (2010) explored the impact, at a food festival, of food neophobia on tourists' satisfaction and loyalty to the festival. They found that food neophobia had a negative effect on satisfaction and loyalty. They insisted that food neophobia should be included as one of the important factors affecting customers' satisfaction and loyalty in tourism and hospitality research. Furthermore, Kim *et al.* (2013) empirically measured tourist food motivation, food-related personality traits and demographic factors. They investigated differences between food neophobia and demographic factors but did not explore the effect of food neophobia on either tourist food motivation or tourist food consumption behaviours.

Given that food neophobia is associated with a fear of consuming novel food, it can greatly influence tourists' food consumption in a destination. However, tourism studies focusing on food neophobia and subsequent investigation into its associations with tourists' socio-demographics and local food experiences in destinations are lacking.

8.2.2 Tourists' local food consumption values and behavioural intention

Consumption value theory explains that consumer choice behaviour is affected by multiple consumption values, such as functional, emotional, social and epistemic values. Each consumption value has a different impact depending on different consumption contexts (Sheth *et al.*, 1991). Consumption value can be defined as the perceived utility obtained from product consumption (Sheth *et al.*, 1991; Sweeney and Soutar, 2001). Consumption value theory has been applied to several tourism studies (Sánchez *et al.*, 2006; Williams and Soutar, 2009; Chen and Chen, 2010). For example, Sánchez *et al.* (2006) applied consumption value theory to measure tourists' consumption values for purchasing a tourism product. They identified functional, emotional and social values in purchasing a tourism package product. They mentioned that consumption values may be different according to different tourism products. Williams and Soutar (2009) examined the relationships among value, satisfaction and behavioural intention in the context of adventure tourism. In their study, functional, emotional and novelty values have been found to influence tourists' satisfaction and future intentions for adventure tours. Choe and Kim (2018) tested the impact of tourists' food consumption values on tourists' attitudes towards local food, image perception of a destination and future behaviours. They found that epistemic, emotional and functional aspects of values are positively related to tourists' favourable attitude towards local food experiences.

Additionally, Choe and Kim (2018) described various types of tourists' local food consumption values. First, taste or quality value is the utility acquired by consuming local food in a destination based on the quality and expected performance of the local food. Second, health value is the utility acquired by consuming local food in a destination based on the ability of the local food to enhance health. Price value is the perceived utility acquired by consuming local food in a destination based on the reduced cost of the local food. Emotional value refers to the perceived utility acquired by consuming local food in a destination based on the capability of the local food to stimulate desired feelings or affective states. Prestige value refers to the perceived utility acquired by consuming local food in a destination based on the capability of the local food to enhance social self-concept. Interaction value is the perceived utility acquired by consuming local food in a destination based on the capability of the local food to enhance interactions among people. Lastly, epistemic value refers to the perceived utility acquired by consuming local food in a destination based on the capability of the local food to arouse curiosity, provide novelty and satisfy one's desire for knowledge (Choe and Kim, 2018).

Behavioural intention is the likelihood that a person engages in specific behaviour (Ajzen and Fishbein, 1980). In many studies, behavioural intention is assumed to become a useful behaviour in the future (Zeithaml *et al.*, 1996). Thus, intention is a critical factor. Identifying the relative importance of the antecedents of tourists' behavioural intention is important because it allows destination marketers to allocate scarce resources efficiently (Prayag and Ryan, 2012). Similarly, tourists' behavioural intention has often been studied in terms of repeat visits and positive recommendation of the destination to others. Keeping pace with the current trend of food tourism, whereby food experiences occur as a significant activity or tourists are motivated to choose a destination to experience local food (Hall *et al.*, 2003), the present study examined tourists' behavioural intention to return to the destination for food tourism.

8.2.3 Differences in socio-demographics, food consumption values and behavioural intention according to FNGs

Previous studies have indicated that people show different levels of food neophobia according to their socio-demographics (Birch, 1999; Tuorila *et al.*, 2001; Kim *et al.*, 2013). For example, among children and adults, males have shown greater food neophobia than females (Koivisto and Sjödén, 1996). According to Tuorila *et al.* (2001), people who showed a lower level of food neophobia were young, highly educated, urbanized women. Birch (1999) demonstrated that food neophobia is minimal during infancy, increases through early childhood and declines from early childhood to adulthood. Mascarello *et al.* (2020) examined the relationships among food neophobia, ethnic food consumption and openness to the different cultures in Italy and attempted to determine who are food neophobic consumers in terms of socio-demographics. They found that food neophobic people are more likely to be male, older, less educated, have children, a low income and a low level of openness to different cultures, and to not eat ethnic food. Although this study was not conducted in the tourism context, it provides strong evidence that different socio-demographic characteristics contribute to forming different levels of food neophobia, which could further influence local food perception and experiences in tourism destinations.

In tourism literature, Cohen and Avieli (2004) mentioned considerable differences among cultures in the extent to which they encourage neophilic tendencies. For example, most Asians have stronger tendencies of food neophobia, whereas westerners are strongly encouraged to try new foods and dishes (Cohen and Avieli, 2004). They also mentioned considerable differences in the levels of food neophobia between social classes and individuals. However, they did not further discuss or suggest the criteria of classification. Furthermore, Kim *et al.*'s (2013) study was the first to explore the relationships between food neophobia and tourists' socio-demographics. They found that males were more neophobic than females. People with higher incomes showed more neophilic tendencies. Previous studies have shown that rising income levels may have driven people to travel and given more opportunity to taste local food in a new destination (Douglas and Derrett, 2001). Similarly, Kim *et al.* (2010) investigated tourists' food neophobia levels at a food festival and found that visitors who were more neophobic were less satisfied with the food festival. Furthermore, tourists who are more neophobic showed a lower likelihood of visiting the festival again. Mak *et al.* (2017) provided evidence of a negative relationship between food neophobia and tourists' food consumption values. Although their study focused on the motivational factors for food in tourism rather than consumption values, they showed that more neophobic tourists were less likely to be motivated by novel foods and a variety of food. Moreover, they found that more neophobic tourists were also interested in interaction, cultural motivations and health.

Choe and Cho (2011) attempted to determine whether food neophobia influences consumers' familiarity and willingness to try ethnic food in Korea. They found significant difference among FNGs (i.e. neophilics, neutrals and neophobics) in the level of familiarity and willingness to try ethnic food, such as Vietnamese, Indian, Thai, Mexican, Turkish and Greek food. However, no significant differences are found among FNGs in the level of familiarity and willingness to try some familiar food for Koreans, such as American, Chinese and Italian food. Similarly, Sogari *et al.* (2019) empirically tested whether food neophobia affected consumers' willingness to eat edible insect products. As a result, more food neophobic consumers showed a low level of acceptance for edible insect products. Wu *et al.* (2016) conducted a multi-group analysis between Chinese tourists with high food neophobia and those with low food neophobia regarding tourist food consumption. As a result, tourists with high and low food neophobia show different psychological reactions and behavioural patterns concerning local food consumption in a destination. Therefore, food neophobia plays an important role in consumers' intention to try new or unfamiliar food. However, only a few studies have identified whether food neophobia influences tourists' intentions to visit the destination to try local delicacies. Based on previous studies, the following hypotheses are proposed. Fig. 8.1 shows the conceptual framework of this study.

8.3 Research Methodology

The ultimate research goal of this study is to see the associations among tourists' food neophobia, socio-demographics and local food experiences in

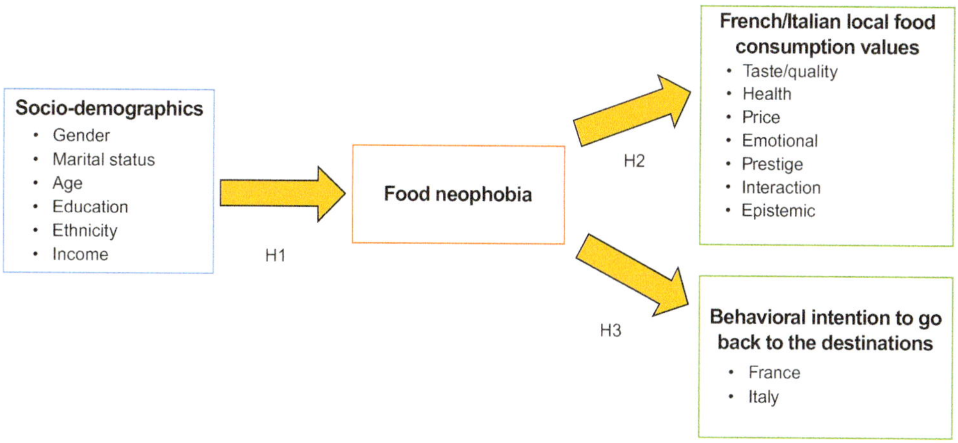

Fig. 8.1.

H1: Socio-demographic differences are found in tourists' levels of food neophobia.

H2: FNGs are likely to show different levels of local food consumption values of tourists.

H3: FNGs are likely to show different levels of behavioural intention.

a destination. Based on the objectives and relevant previous studies, the decision has been made to employ quantitative research methods. Quantitative results can be summarized numerically with a certain level of confidence (Field, 2013). Quantitative methods are believed to be effective in terms of representation (Creswell, 2013).

8.3.1 Questionnaire design

The questionnaire consisted of three parts. The first part contained six statements about food neophobia (Pliner and Hobden, 1992; Ritchey *et al.*, 2003). The second section consisted of 30 and 3 questions to assess tourists' consumption values (Choe and Kim, 2018) and behavioural intention to visit the destination for food tourism (Kim *et al.*, 2014), respectively. The last part consisted of socio-demographic and travel-related questions. Measurement items of the main constructs used in this study are shown in Appendix 1. A seven-point Likert-type scale (1=strongly disagree, 7=strongly agree) was used for food neophobia, consumption values and behavioural intention.

8.3.2 Data and sample

Through Qualtrics, an online survey company in the USA, an online questionnaire was made and distributed, and the data were collected in 2016.

Qualtrics initially sent invitation e-mails to 3000 panels, of which 420 took part in the survey. After checking multi-variate outliers and visual inspections, all the responses were used. The target population for the study was composed of US tourists who had travelled to France or Italy in the past two years.

US tourists were chosen among English-speaking tourists because they constitute one of the top outbound tourist segments in the world (UNWTO, 2016). The top two tourism destinations for US citizens are Mexico and Canada (Office of Travel and Tourism Industries, 2015; TravelBizNews, 2018). However, Mexico and Canada were excluded as overseas travel destinations in this study because many travellers cross the borders from the USA to Mexico and Canada frequently for various reasons other than leisure or vacation. Apart from Mexico and Canada, western Europe is the most popular region for US tourists (The Telegraph, 2018). France and Italy were selected as the destinations because they are two of US tourists' top outbound destinations (Office of Travel and Tourism Industries, 2015) and have rich unique food cultures and cuisines. According to Ab Karim and Chi (2010), France and Italy have positive culinary destination images in terms of their long history and fine European cuisines. These two countries have a food destination image and provide tasty, presentative, varied and quality food to tourists. Vujicic *et al.* (2013) conducted a multi-country survey on foodies and food tourism by asking people to rank the best and the second-best food destination image. In their study, respondents pointed out that Italy was the best and France was the second-best destination for food.

8.3.3 Food Neophobia Scale (FNS) and data analysis

FNS scores are calculated by adding the individual scores for each item (Pliner and Hobden, 1992). The FNS scores range from 6 to 35 for French food and 6 to 33 for Italian food; a higher value designates a higher level of food neophobia. The respondents were divided into tertials representing low, medium and high FNGs (i.e. neophilics, neutrals and neophobics) based on the grand mean and standard deviation to see any relationships clearly between the participants at each end of the FNS. This type of classification has been used in previous studies (Tuorila *et al.*, 2001; Ritchey *et al.*, 2003; Olabi *et al.*, 2009; Choe and Cho, 2011) and is considered to be a corroborative method. For data analysis, t-test and analysis of variance (ANOVA) using IBM SPSS Statistics 24 were performed to examine the effect of food neophobia on the outcome variables.

8.4 Findings and Discussions

Table 8.1 shows the profile of the respondents. The demographic characteristics of the respondents for this study have seven components, namely gender, age, education level, ethnic background, marital status, occupation

Table 8.1. The socio-demographics of the respondents.

US tourists who have travelled to France (*N* = 210)			US tourists who have travelled to Italy (*N* = 210)		
Category		Percentage (%)	Category		Percentage (%)
Gender	Female	54.4	Gender	Female	53.9
	Male	45.6		Male	46.1
Age	20s	21.8	Age	20s	27.7
	30s	33.0		30s	29.1
	40s	13.1		40s	12.1
	50s	12.6		50s	12.6
	60s	9.7		60s	13.6
	70s or above	9.7		70s or above	4.9
Educational level	High school or less	11.2	Educational level	High school or less	11.7
	College student	13.1		College student	11.2
	College degree	52.4		College degree	50.0
	Master's degree	20.9		Master's degree	24.8
	Doctoral degree	2.4		Doctoral degree	2.4
Ethnic background	Caucasian	79.1	Ethnic background	Caucasian	78.2
	African American	5.8		African American	4.4
	Asian	3.9		Asian	6.3
	Hispanic	7.8		Hispanic	9.7
	Others	3.4		Others	1.5
Marital status	Single	34.5	Marital status	Single	35.9
	Married	62.1		Married	60.7
	Others	3.4		Others	3.4
Occupation	Company employee	26.7	Occupation	Company employee	26.7
	Own business	10.7		Own business	9.7
	Civil servant	1.5		Civil servant	1.0
	Professional	15.5		Professional	15.5
	Housewife	7.3		Housewife	4.4
	Technician	0.5		Technician	2.9
	Student	4.9		Student	4.9
	Sales/service	6.3		Sales/service	0.5
	Education	4.4		Education	8.7
	Retired	14.6		Retired	14.6
	Others	7.8		Others	11.2

Continued

Table 8.1. Continued.

US tourists who have travelled to France (*N* = 210)			US tourists who have travelled to Italy (*N* = 210)		
Category		Percentage (%)	Category		Percentage (%)
Monthly household income (US$)	Less than 1000	3.4	Monthly household income (US$)	Less than 1000	1.0
	1001–2000	9.2		1001–2000	13.1
	2001–3000	13.1		2001–3000	19.4
	3001–4000	18.4		3001–4000	18.9
	4001–5000	17.5		4001–5000	13.6
	5001–6000	14.1		5001–6000	9.7
	6001–7000	8.3		6001–7000	6.3
	7001–8000	8.3		7001–8000	5.8
	8001 or above	7.8		8001 or above	12.1

and monthly household income. The percentage of female US tourists who have travelled to France was 54.4. Those in their 30s had the highest percentage (33.0%). The majority of the respondents had a college degree (52.4%). Most of the respondents were Caucasian (79.1%). The percentage of married respondents was 62.1. Approximately 26.7% of the respondents were company employees, and 15.5% were professionals. Meanwhile, 18.4% of the respondents reported earnings of US$3001–4000. The demographics of US tourists who have travelled to Italy were similar to those who have travelled to France. The percentage of females was slightly higher at 53.9. The respondents in their 30s comprised the majority of the respondents (29.1%). The majority of the respondents had a Master's degree (24.8%). Caucasian respondents comprised the majority (78.2%). Approximately 26.7% of the respondents were company employees, whereas 15.5% were professionals. Finally, 18.9% of the respondents earned US$ 3001–4000.

Table 8.2 shows FNGs for French and Italian food experiences. Respondents were classified into groups representing low food neophobia for scores between 6 and 9, medium food neophobia for scores between 10 and 21 and high food neophobia for scores between 22 and 35, in the case of French food. Similarly, tourists were classified into groups. Those representing low food neophobia have scores between 6 and 9, medium food neophobia have scores between 10 and 21 and high food neophobia have scores between 22 and 33, in the case of Italian food. As a result, the respective percentages of subjects in each sub-group was 20.5 (neophilics), 65.2 (neutrals) and 14.3 (neophobics) for French food experiences and 22.4 (neophilics), 64.8 (neutrals) and 12.9 (neophobics) for Italian food experiences.

Table 8.2. Food neophobia groups ($N = 420$).

FNGs		N (%)
French food experience (N = 210)	Neophilics	43 (20.5)
	Neutrals	137 (65.2)
	Neophobics	30 (14.3)
Italian food experience (N = 210)	Neophilics	47 (22.4)
	Neutrals	136 (64.8)
	Neophobics	27 (12.9)

The results of the t-test showed statistically significant differences between the female and male group ($p < 0.05$) and the single and married group ($p < 0.05$) for the food neophobia mean score. Males were found to be more neophobic than females. Furthermore, married people are more neophobic. These findings are consistent with some previous studies that indicated that neophobic consumers are likely to be male and married with children (Tuorila *et al.*, 1998; Kim *et al.*, 2013; Mascarello *et al.*, 2020).

The results of the ANOVA test indicated statistically significant differences between age groups ($p < 0.05$) and ethnicity groups ($p < 0.05$) for the food neophobia mean score. Younger and older people were more neophobic than middle-aged people. This result is different from previous studies (e.g. Tuorila *et al.*, 2001), which have indicated that older people are more neophobic than younger people. Cohen and Avieli (2004) stated that westerners are more likely to be neophilic. However, in our study, Caucasians' food neophobia levels were the highest. No differences were observed in the food neophobia levels between different education and income levels. This result is different from our expectations. Previous studies (e.g. Tuorila *et al.*, 2001) have stated that high income and education gives more opportunities for people to be exposed to diverse food stimuli, activities and events, decreasing food neophobia.

Furthermore, tourists who indicated that experiencing French/Italian food was one of their major motivations for visiting the destination and one of the most important parts of their trip were significantly less neophobic. Neophilic and neutral tourists were found to buy French/Italian food as a gift or souvenir. The results of this current study show that hypothesis 1 was partially supported (see Table 8.3).

Table 8.4 and Fig. 8.2 show the main effects of FNGs on taste/quality value, health value, price value, emotional value, prestige value, interaction value, epistemic value and behavioural intention to return to the destinations for food tourism. The results for ANOVA regarding the effects of FNGs on tourists' French food consumption values and their behavioural intention indicated statistically significant differences among FNGs ($p < 0.05$) for taste or quality value, health value, price value, emotional value, prestige value, interaction value, epistemic value and behavioural intention to go back to the destination for food tourism. Similarly, the results for ANOVA regarding the effects of FNGs on tourists' Italian food consumption values and their behavioural intention indicated statistically significant differences among FNGs

Table 8.3. The results of the independent samples t-test and ANOVA regarding the socio-demographic differences of food neophobia.

	Category	Mean	t-value /F-value
Gender	Female (*N* = 226)	14.44	–2.77*
	Male (*N* = 194)	16.10	
Marital status	Single (*N* = 163)	14.36	–2.24*
	Married (*N* = 257)	15.74	
Age	Younger (20s) (*N* = 104)	16.27a	8.020***
	Middle (30–40s) (=183)	13.58b	
	Older (50s or above) (*N* = 133)	16.24a	
Education	Below than college (*N* = 102)	14.90	1.861
	College degree (*N* = 202)	14.85	
	Master's and doctoral degree(*N* = 106)	16.21	
Ethnicity	Caucasian (*N* = 330)	15.71	2.971*
	African American (*N* = 21)	14.00	
	Asian (*N* = 21)	12.00	
	Hispanic (*N* = 38)	13.49	
	Others (*N* = 10)	14.30	
Monthly household income	Less than US$39,999 (*N* = 57)	15.60	0.211
	US$40,000–99,999 (*N* = 209)	15.26	
	Higher than US$100,000 (*N* = 154)	14.99	

*p < .05, ***p < .001

($p < 0.05$) for taste/quality value, health value, price value, emotional value, prestige value, interaction value, epistemic value and behavioural intention to go back to the destination for food tourism. This result is similar to that of previous studies that found that food neophobia has a negative influence on tourists' local food motivations (Mak *et al.*, 2017; Gurbaskan Akyuz, 2019). However, previous studies have focused on the investigation between food neophobia and tourists' food motivation, which is different from tourists' perceived utility of local food consumption.

In sum, all the outcome variables showed significant differences among the FNGs. More food neophobic tourists perceived a low level of food consumption values. More neophobic tourists showed a lower level of behavioural intention to return to the destinations. The results of the French and Italian food cases were similar. Hypotheses 2 and 3 were supported.

8.5 Theoretical and Practical Implications

The objectives of this study are to explore whether tourists have different levels of food neophobia according to their socio-demographics and to examine the relationships between different food neophobia groups and their local food consumption values and behavioural intention to go back

Table 8.4. The results for ANOVA regarding the effects of FNGs on tourists' food consumption values and their behavioural intention.

	French food experiences (*N* = 210)				Italian food experiences (*N* = 210)			
	Neophilics	Neutrals	Neophobics	F-value	Neophilics	Neutrals	Neophobics	F-value
Taste/quality value	6.59a	6.19a	5.43b	14.53***	6.87a	6.42b	5.60c	28.86***
Health value	6.02a	5.90a	5.01b	11.58***	6.55a	6.01b	5.07c	29.02***
Price value	5.81a	5.57a	4.68b	7.64**	6.54a	6.02b	5.20c	18.89***
Emotional value	6.38a	5.87b	5.16c	13.41***	6.55a	6.10b	5.01c	29.76***
Prestige value	5.35a	5.07ab	4.42b	3.58*	5.66a	5.33b	4.40b	10.03***
Interaction value	6.40a	5.71b	5.31b	10.84***	6.42a	5.85b	4.85c	25.53***
Epistemic value	6.31a	5.80b	5.57b	7.86**	6.47a	6.00b	5.04c	25.15***
Behavioural intention to go back to the destination for food tourism	6.54a	5.89b	4.83c	18.47***	6.65a	6.09b	4.71c	32.75***

*p < .05, **p < .01, ***p < .001 ; a, b, and c indicate the source of significant difference (a < b < c); Post-hoc test results are based on Tukey HSD and Duncan.

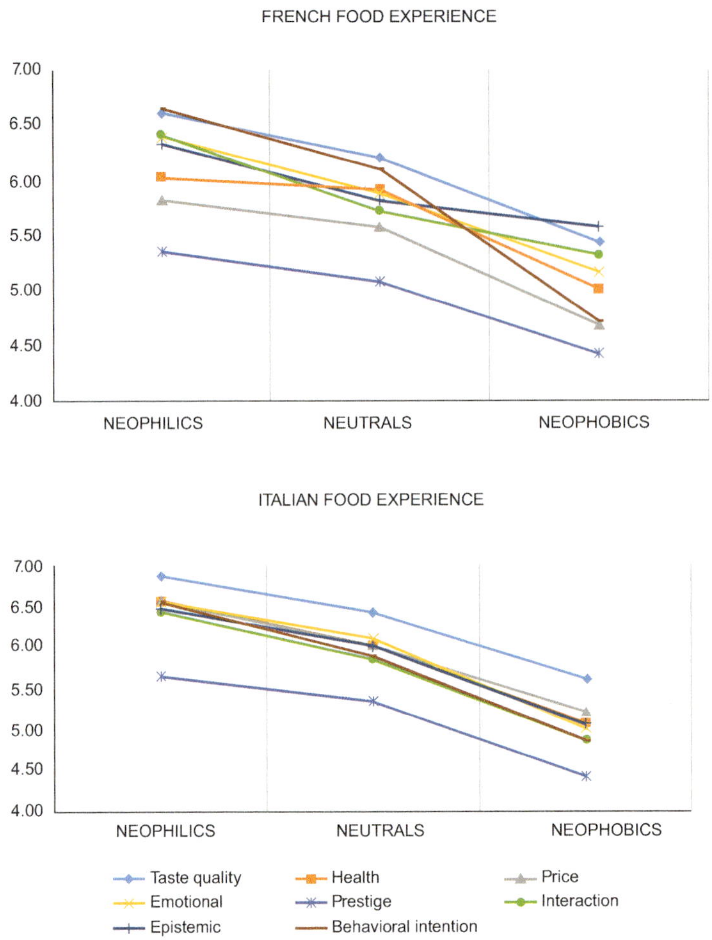

Fig. 8.2. The mean ratings of tourists' local food consumption values and behavioural intention in the three FNGs (neophilics, neutrals and neophobics).

to the destination for food tourism. The results of this study show that food neophobia in different socio-demographic groups needs further attention and should also be tested in other contexts because some results were different from previous studies. Moreover, the level of tourists' local food consumption values and behavioural intention was lower than the more neophobic group. That is, if tourists are food-neophobic, they may consider local food that is less tasty and healthy is less good value for money, is less likely to make thems happy or excited, and provides social values and new knowledge about the local food culture. Moreover, if tourists are more neophobic, they are less likely to come back to the destination to try local delicacies.

From a theoretical perspective, the current study broadens the range of studies regarding food neophobia in tourism. Despite its importance, food neophobia was not fully researched in tourism literature. Many previous studies have conceptually suggested the importance of knowing food neophobia to understand tourist food consumption better. However, empirical studies that have attempted to understand the influence of food neophobia on tourists' perceived utilities of local food experiences and future intention were limited. This study also contributes to the first empirical application of food neophobia concept and the consumption value theory to the context of tourists' local food consumption. Furthermore, this study confirmed that the fear of novel food formed daily has a negative impact on their local food experiences in a destination.

From a practical point of view, destination marketers, hospitality businesses and tour operators can achieve a competitive advantage by understanding the concept of food neophobia. Destination marketers should consider food neophobia when they plan and design food-related activities for tourists. Particularly, this study provides insight into French and Italian destination food marketers for their planning to attract US tourists. More neophobic tourists from the USA are found to be male, married and Caucasian. Female, single, middle-aged and non-Caucasian US tourists are targeted first when it comes to promoting unique and exotic local cuisines to attract people with high neophilic tendencies.

However, particular attention should also be paid to food neophobic segments. Lai *et al.* (2020) suggested that a dual strategy for tourists with high and low neophobia is necessary. The current study also confirms the importance of the argument suggested by Lai *et al.* (2020) because two different groups (high and low FNGs) have a different psychological mindset, attitude and behavioural intention towards unfamiliar local food in a destination. Based on the current findings, previous studies and personal insights, several practical suggestions have been made in the following sections.

First, while promoting local delicacies, making US neophobic tourists feel more comfortable when they consume local food in France and Italy is important. For example, although the food itself can be new, the presentation of dishes and table setting can be adjusted to an international or US style for more neophobic tourists. Second, tourism labels that guarantee the quality of local food should be more emphasized in restaurants to make neophobic tourists feel more relieved. For example, information about Qualité Tourisme and Le Cordon Bleu restaurants in France can be more emphasized to food-neophobic tourists. Third, giving the impression to food-neophobic tourists that they can 'control' and 'choose' what they want to eat is important because reducing tourists' feelings of unfamiliarity and anxiety regarding local cuisines in a destination is important. According to Wu *et al.* (2016), more food-neophobic tourists who think that they can control their unfamiliar local food consumption during travelling showed a high level of intention to try unfamiliar local food. Thus, French and Italian local restaurants can have a variety of options for tourists. For example, French and Italian local restaurants can provide different types of menus with more international tastes to high FNGs.

Fourth, local cuisines can be offered together with some familiar food to neophobic tourists. For example, 'pani ca meusa' (bun stuffed with veal lung and spleen) from Italy or 'lapin au vin blanc' (rabbit meat in wine) from France can be served as an American-style burger to reduce unfamiliar and uncomfortable feelings. Fifth, previous studies from the food science and consumer research fields have indicated that food neophobia is only reduced when people learn that the novel food is safe to eat and does not cause illness (Kalat and Rozin, 1973). In tourism literature, Wu *et al.* (2016) also emphasized that restaurants must ensure the safety of the food. If tourists cannot trust a safe and hygienic environment, appealing to tourists, especially to people who are more food neophobic, is difficult. Thus, we particularly recommend that French and Italian destination marketers focus on building trust among US tourists and making sure that the local cuisine is safe to try especially for food neophobic people. Providing an English menu with a detailed description will be helpful for US tourists to understand the local cuisines in a destination. Although emphasizing the authenticity, uniqueness and exotic flavour of local cuisines to low-food-neophobc tourists was recommended, many have suggested focusing on the similarities of food ingredients, sauces and the cooking methods used between two countries to achieve a more neophobic group.

Last, through child behaviour studies, five and ten exposures to a novel food were necessary before seeing an increased food preference for the novel food in children (Birch, 1999). Similarly, Tuorila *et al.* (2001) mentioned that willingness to try was enhanced significantly by earlier tasting. Particularly, tasting experience was beneficial for high-neophobic people. According to Tuorila *et al.* (1994), the tasting experience helps individuals to recognize whether the sensory attributes of food are better or worse than their expectation. Therefore, increasing the number of exposures of local food to US tourists will be helpful because previous research has indicated that the neophobic response is reduced by repeated opportunities to consume new foods. Thus, providing opportunities to sample good-tasting novel local foods is necessary. For example, France and Italy can organize their food-related workshops, media events and various food festivals in the US to appeal to potential US tourists. These types of activities can increase the knowledge of neophilic and neophobic tourists from the USA about French and Italian food culture and enhance their positive expectation of trying authentic local food in the destination. That is, providing more opportunities for neophobic tourists to be exposed to diverse ethical foods (e.g. French and Italian food) in a familiar environment (e.g. hometown) contributes to decreasing their fear of novel food.

8.6 Conclusions and Future Research

The current study is not free from limitations. First, a narrow sample of participants is one of the limitations. In future studies, larger sample size can be applied to measure food neophobia and tourists' local food consumption values and behavioural intentions. Second, this study mainly adopted the quantitative research method but, in the future, a mixed-method approach can provide insightful information on tourists' food-related personality traits and local food consumption in a destination. Interviewing tourists about the different food neophobia levels, how they deal with local food consumption and how their psychological and behavioural patterns change regarding food consumption in a destination would be interesting. Third, this study investigated US tourists' local food experiences in France and Italy, but this can be expanded to diverse contexts with different samples. For example, more cross-cultural research into food neophobia in different destinations with tourists from diverse backgrounds can be conducted.

To conclude, when destination marketers are attempting to attract tourists with neophilic tendencies, marketers are suggested to emphasize authentic cuisine, exotic dishes and special spices that represent the local identity. However, destination marketers should also remember to focus on how their non-traditional foods can be made more adaptable and familiar to more neophobic tourists who are also in an important market by utilizing the above-mentioned suggestions. By understanding the concept of food neophobia, practitioners can gain a competitive advantage in the tourism and hospitality industry.

References

Ab Karim, S. and Chi, C.G.Q. (2010) Culinary tourism as a destination attraction: an empirical examination of destinations' food image. *Journal of Hospitality Marketing and Management* 19(6), 531–555. DOI: 10.1080/19368623.2010.493064.

Ajzen, I. and Fishbein, M. (1980) *Understanding Attitudes and Predicting Social Behaviour*. Prentice-Hall, Englewood Cliffs, New Jersey.

Birch, L.L. (1999) Development of food preferences. *Annual Review of Nutrition* 19, 41–62. DOI: 10.1146/annurev.nutr.19.1.41.

Chang, R.C., Kivela, J. and Mak, A.H.N. (2011) Attributes that influence the evaluation of travel dining experience: when east meets west. *Tourism Management* 32(2), 307–316. DOI: 10.1016/j.tourman.2010.02.009.

Chen, C. and Chen, F. (2010) Experience quality, perceived value, satisfaction and behavioural intentions for heritage tourists. *Tourism Management* 31(1), 29–35. DOI: 10.1016/j.tourman.2009.02.008.

Choe, J.Y.J. and Cho, M.S. (2011) Food neophobia and willingness to try non-traditional foods for Koreans. *Food Quality and Preference* 22(7), 671–677. DOI: 10.1016/j.foodqual.2011.05.002.

Choe, J.Y.J. and Kim, S.S. (2018) Effects of tourists' local food consumption value on attitude, food destination image, and behavioural intention. *International Journal of Hospitality Management* 71(4), 1–10. DOI: 10.1016/j.ijhm.2017.11.007.

Cohen, E. (1972) Toward a sociology of international tourism. *Social Research* 39, 164–182.

Cohen, E. and Avieli, N. (2004) Food in tourism: attraction and impediment. *Annals of Tourism Research* 31, 755–778.

Creswell, J.W. (2013) *Research Design: Qualitative, Quantitative, and Mixed Methods Approaches*. Sage Publications, Thousand Oaks, California.

Douglas, N. and Derrett, R. (2001) *Special Interest Tourism*. John Wiley and Sons Australia, Ltd, Milton.

Field, A. (2013) *Discovering Statistics Using IBM SPSS Statistics*. Sage, London.

Fields, K. (2002) Demand for the gastronomy tourism product: motivational factors. In: Hjalager, A.M. and Richards, G. (eds) *Tourism and Gastronomy*. Routledge, London, pp. 37–50.

Fischler, C. (1988) Food, self and identity. *Information (International Social Science Council)* 27(2), 275–292. DOI: 10.1177/053901888027002005.

Getz, D. and Robinson, R.N.S. (2014) Foodies and food events. *Scandinavian Journal of Hospitality and Tourism* 14(3), 315–330. DOI: 10.1080/15022250.2014.946227.

Gurbaskan Akyuz, B. (2019) Factors that influence local food consumption motivation and its effects on travel intentions. *Anatolia* 30(3), 358–367. DOI: 10.1080/13032917.2019.1595072.

Hall, C.M., Sharples, E., Mitchell, R., Cambourne, B. and Macionis, N. (eds) (2003) *Food Tourism Around the World: Development, Management and Markets*. Butterworth-Heinemann, Oxford, UK.

Henriques, A.S., King, S.C. and Meiselman, H.L. (2009) Consumer segmentation based on food neophobia and its application to product development. *Food Quality and Preference* 20(2), 83–91. DOI: 10.1016/j.foodqual.2008.01.003.

Kalat, J.W. and Rozin, P. (1973) 'Learned safety' as a mechanism in long-delay taste-aversion learning in rats. *Journal of Comparative and Physiological Psychology* 83(2), 198–207. DOI: 10.1037/h0034424.

Kim, Y.G., Suh, B.W. and Eves, A. (2010) The relationships between food-related personality traits, satisfaction, and loyalty among visitors attending food events and festivals. *International Journal of Hospitality Management* 29(2), 216–226. DOI: 10.1016/j.ijhm.2009.10.015.

Kim, Y.G., Eves, A. and Scarles, C. (2013) Empirical verification of a conceptual model of local food consumption at a tourist destination. *International Journal of Hospitality Management* 33(3), 484–489. DOI: 10.1016/j.ijhm.2012.06.005.

Kim, S.S., Agrusa, J. and Chon, K. (2014) The influence of a TV drama on visitors' perception: a cross-cultural study. *Journal of Travel and Tourism Marketing* 31(4), 536–562. DOI: 10.1080/10548408.2014.883950.

Koivisto, U.K. and Sjödén, P.O. (1996) Food and general neophobia in Swedish families: parent–child comparisons and relationships with serving specific foods. *Appetite* 26(2), 107–118. DOI: 10.1006/appe.1996.0009.

Lai, M.Y., Wang, Y. and Khoo-Lattimore, C. (2020) Do food image and food neophobia affect tourist intention to visit a destination? The case of Australia. *Journal of Travel Research* 59(5), 928–949. DOI: 10.1177/0047287519867144.

Long, L.M. (2004) *Culinary Tourism*. University Press of Kentucky, Lexington, Kentucky.

Mak, A.H., Lumbers, M. and Eves, A. (2012) Globalization and food consumption in tourism. *Annals of Tourism Research* 39(1), 171–196. DOI: 10.1016/j.annals.2011.05.010.

Mak, A.H., Lumbers, M., Eves, A. and Chang, R.C. (2017) The effects of food-related personality traits on tourist food consumption motivations. *Asia Pacific Journal of Tourism Research* 22(1), 1–20. DOI: 10.1080/10941665.2016.1175488.

Mascarello, G., Pinto, A., Rizzoli, V., Tiozzo, B., Crovato, S. *et al.* (2020) Ethnic food consumption in Italy: the role of food neophobia and openness to different cultures. *Foods* 9(2), 112. DOI: 10.3390/foods9020112.

Mkono, M., Markwell, K. and Wilson, E. (2013) Applying Quan and Wang's structural model of the tourist experience: a Zimbabwean netnography of food tourism. *Tourism Management Perspectives* 5(3), 68–74. DOI: 10.1016/j.tmp.2012.10.007.

Molz, G.J. (2007) Eating difference: the cosmopolitan mobilities of culinary tourism. *Space and Culture* 10, 77–93.

Nordin, S., Broman, D.A., Garvill, J. and Nyroos, M. (2004) Gender differences in factors affecting rejection of food in healthy young Swedish adults. *Appetite* 43(3), 295–301. DOI: 10.1016/j.appet.2004.07.002.

Office of Travel and Tourism Industries (2015) U.S. resident ravel to international destinations increased 10 percent in 2014. Available at: http://travel.trade.gov/research/monthly/departures/ (accessed 18 January 2021).

Olabi, A., Najm, N.E.O., Baghdadi, O.K. and Morton, J.M. (2009) Food neophobia levels of Lebanese and American college students. *Food Quality and Preference* 20(5), 353–362. DOI: 10.1016/j.foodqual.2009.01.005.

Petrick, J.F. and Backman, S.J. (2002) An examination of the construct of perceived value for the prediction of golf travelers' intentions to revisit. *Journal of Travel Research* 41(1), 38–45. DOI: 10.1177/004728750204100106.

Pliner, P. and Hobden, K. (1992) Development of a scale to measure the trait of food neophobia in humans. *Appetite* 19(2), 105–120. DOI: 10.1016/0195-6663(92)90014-W.

Pliner, P. and Salvy, S.J. (2006) Food neophobia in humans. In: Shepherd, R. and Raats, M. (eds) *The Psychology of Food Choice*. CAB International, Wallingford, UK, pp. 75–92.

Prayag, G. and Ryan, C. (2012) Antecedents of tourists' loyalty to Mauritius: the role and influence of destination image, place attachment, personal involvement, and satisfaction. *Journal of Travel Research* 51, 342–356.

Ritchey, P.N., Frank, R.A., Hursti, U.K. and Tuorila, H. (2003) Validation and cross-national comparison of the food neophobia scale (FNS) using confirmatory factor analysis. *Appetite* 40(2), 163–173. DOI: 10.1016/S0195-6663(02)00134-4.

Rozin, P. (1976) The selection of food by rats, humans and other animals. In: Rosenblatt, J., Hinde, R.A., Beer, C. and Shaw, E. (eds) *Advances in the Study of Behaviour*. Academic Press, San Diego, California, pp. 21–76.

Ryu, K., Han, H. and Kim, T. (2008) The relationships among overall quick-casual restaurant image, perceived value, customer satisfaction, and behavioural intentions. *International Journal of Hospitality Management* 27(3), 459–469. DOI: 10.1016/j.ijhm.2007.11.001.

Sánchez, J., Callarisa, L., Rodríguez, R.M. and Moliner, M.A. (2006) Perceived value of the purchase of a tourism product. *Tourism Management* 27(3), 394–409. DOI: 10.1016/j.tourman.2004.11.007.

Sheth, J.N., Newman, B.I. and Gross, B.L. (1991) *Consumption Values and Market Choices: Theory and Applications*. South-Western Publishing, Cincinnati, Ohio.

Siegrist, M., Hartmann, C. and Keller, C. (2013) Antecedents of food neophobia and its association with eating behaviour and food choices. *Food Quality and Preference* 30(2), 293–298. DOI: 10.1016/j.foodqual.2013.06.013.

Sogari, G., Menozzi, D. and Mora, C. (2019) The food neophobia scale and young adults' intention to eat insect products. *International Journal of Consumer Studies* 43(1), 68–76. DOI: 10.1111/ijcs.12485.

Stanley, J. and Stanley, L. (2014) *Food Tourism: A Practical Marketing Guide*. CAB International, Wallingford, UK.

Sweeney, J.C. and Soutar, G.N. (2001) Consumer perceived value: the development of a multiple item scale. *Journal of Retailing* 77(2), 203–220. DOI: 10.1016/S0022-4359(01)00041-0.

The Telegraph (2018) The truth about how Americans travel. Available at: https://www.telegraph.co.uk/travel/destinations/north-america/united-states/articles/what-american-tourists-are-really-like/ (accessed 5 July 2019).

Torres, R. (2002) Toward a better understanding of tourism and agriculture linkages in the Yucatan: tourist food consumption and preferences. *Tourism Geographies* 4(3), 282–306. DOI: 10.1080/14616680210147436.

TravelBizNews (2018) U.S. international travel up in 2017, outbound travel 80.2 million last year. Available at: https://travelbiznews.com/u-s-international-travel-up-in-2017-outbound-travel-80-2-million-last-year/ (accessed 2 February 2021).

Tuorila, H., Meiselman, H.L., Bell, R., Cardello, A.V. and Johnson, W. (1994) Role of sensory and cognitive information in the enhancement of certainty and linking for novel and familiar foods. *Appetite* 23(3), 231–246. DOI: 10.1006/appe.1994.1056.

Tuorila, H., Meiselman, H.L., Cardello, A.V. and Lesher, L.L. (1998) Effect of expectations and the definition of product category on the acceptance of unfamiliar foods. *Food Quality and Preference* 9(6), 421–430. DOI: 10.1016/S0950-3293(98)00012-3.

Tuorila, H., Lähteenmäki, L., Pohjalainen, L. and Lotti, L. (2001) Food neophobia among the Finns and related responses to familiar and unfamiliar foods. *Food Quality and Preference* 12(1), 29–37. DOI: 10.1016/S0950-3293(00)00025-2.

UNWTO (2016) UNWTO Tourism Highlights. Available at: http://www.e-unwto.org/doi/pdf/10.18111/9789284418145 (accessed 2 February 2021).

Vujicic, S., Getz, D. and Robinson, R.N.S. (2013) *Food tourists: who are they, what do they want, how do you package offers, and reach them?* Goteborg and Co., Goteborg, Sweden.

Williams, P. and Soutar, G.N. (2009) Value, satisfaction and behavioural intentions in an adventure tourism context. *Annals of Tourism Research* 36(3), 413–438. DOI: 10.1016/j.annals.2009.02.002.

Wu, K., Raab, C., Chang, W. and Krishen, A. (2016) Understanding Chinese tourists' food consumption in the United States. *Journal of Business Research* 69(10), 4706–4713. DOI: 10.1016/j.jbusres.2016.04.018.

Zachmann, K. and Østby, P. (2011) Food, technology, and trust: an introduction. *History and Technology* 27(1), 1–10. DOI: 10.1080/07341512.2011.548970.

Zeithaml, V.A., Berry, L.L. and Parasuraman, A. (1996) The behavioural consequences of service quality. *Journal of Marketing* 60(2), 31–46. DOI: 10.1177/002224299606000203.

Appendix 1 – Measurement Items

Constructs	Measurement items
Food neophobia	I constantly taste new and different food (reversed).
	If I do not know what is in a dish, I will not try it.
	I like food from different countries (reversed).
	At dinner parties, I will try new food (reversed).
	I am afraid to eat things that I have never had before.
	I like to try a new ethnic restaurant (reversed).
Taste/quality value	Overall, I think that French/Italian food was tasty to me.
	Overall, I think that French/Italian food provided appealing flavours.
	Overall, I think that French/Italian food provided a variety of ingredients.
	Overall, I think that French/Italian food provided good-quality ingredients.
	Overall, I think that French/Italian food provided a high standard of quality.
Health value	Overall, I think that French/Italian food provided good nutrition.
	Overall, I think that French/Italian food made me healthy.
	Overall, I think that French/Italian food was hygienic.
	Overall, I think that French/Italian food was safe.
Price value	Overall, I think that French/Italian food was reasonably priced.
	Overall, I think that French/Italian food offered value for money.
Emotional value	Overall, I think that eating French/Italian food made me feel excited.
	Overall, I think that eating French/Italian food gave me pleasure.
	Overall, I think that eating French/Italian food made me feel happy.
	Overall, I think that eating French/Italian food changed my mood positively.
	Overall, I think that French/Italian food made me crave it.
	Overall, I think that I was fascinated by eating French /Italian food.
Prestige value	Overall, I think that eating well-known French/Italian food gave me prestige.
	Overall, I think that I had higher social status when eating well-known French or Italian food.
	Overall, I think that I belonged to one of many people who already experienced French/Italian food.
	Overall, I think that it was worthy of showing pictures of the French/Italian food experiences to others.
	Overall, I think that eating French food gave me a chance to show off my French/Italian food experiences to others.
Interaction value	Overall, I think that eating French/Italian food helped me interact with local people.
	Overall, I think that eating French/Italian food helped me interact with the people I travel with.
	Overall, I think that my friendship or kinship with the travel companion increased while eating French/Italian food together.

Continued

Constructs	Measurement items
Epistemic value	Overall, I think that I satisfied my curiosity by exploring French/Italian food.
	Overall, I think that I experienced French/Italian culture by eating their food.
	Overall, I think that I learned French/Italian dining habits through my French food experiences (e.g. how to eat the food, how to use utensils).
	Overall, I think that my knowledge about French/Italian culture increased.
	Overall, I think that eating French/Italian food was good experience for me to learn new things.
Behavioural intentions for food tourism	I would like to go back to France/Italy to enjoy French food within the next three years.
	I would like to travel to France/Italy for food tourism within the next three years.
	I would like to revisit France/Italy to explore diverse French/Italian foods within the next three years.

9 Nazi Past and Destination Image: The Case of Linz, Austria

Elitza Iordanova*

University of West London, London Geller College of Hospitality and Tourism, Ealing, London, UK

9.1 Introduction

Over the past decades, many places believed they had found a 'saviour' in the form of tourism development as a major source of economic benefit, along with improved local communities' quality of life (Dumont *et al.*, 2010). In this competitive environment, visuality is a key issue as destination management organizations (DMOs) invest much effort and resources in creating an image that will establish a destination as attractive in tourists' minds (Dumont *et al.*, 2010; Kneesel *et al.*, 2010) in an environment where substitutability is becoming more prominent (Agapito *et al.*, 2010). Furthermore, due to the intangible nature of the tourism product, travellers are believed to base their buying decisions on the mental images they have of places (Buhalis, 2000; Chen and Tsai, 2007). This intangibility makes destination image a major marketing tool in the tourism industry (Tasci and Kozak, 2006) used by DMOs to achieve competitive advantage (Konecnik, 2002; Agapito *et al.*, 2010; Ritchie and Crouch, 2010; Fernando and Long, 2012) and to differentiate themselves from their competitors (Echtner and Ritchie, 1993; Carballo *et al.*, 2015). Nevertheless, destinations burdened with crime, wars, natural disasters etc., or difficult heritage, see their competitive advantage diminishing in the eyes of potential tourists (Crouch and Ritchie, 1999). There is also a plethora of tourism studies indicating that destination image is a direct antecedent of satisfaction (Bigné *et al.*, 2001; Chen and Tsai, 2007; Qu *et al.*, 2011; Chen and Phou, 2013; Lee *et al.*, 2014; Alcocer and Ruiz, 2020) and the more positive the image is, the greater the level of satisfaction (Chi and Qu, 2008). This could lead to repeat visits (Kozak, 2001). Moreover, it is acknowledged that only destinations with strong and positive images have a chance of being

*elitza.Iordanova@uwl.ac.uk

DOI: 10.1079/9781789248753.0009

evaluated and, consequently, chosen by tourists (Chen and Kerstetter, 1999; Pan, 2011; Ahn *et al.*, 2013), which could then be further positively modified as a result of actual visitations (Baloglu, 2001).

Destinations, however, tend to use the same destination image to attract visitors regardless of their socio-demographic characteristics, including country of origin (Bonn *et al.*, 2005), which has led to the notion in the literature that DMOs need to 'acknowledge the fact that the same destination may mean different things to different stakeholders' (Stylidis *et al.*, 2015, p. 712). This statement corresponds with Mayo and Jarvis (1981) that 'no two people see a destination in exactly the same way' (p. 42). Nevertheless, any differences in perceptions among various stakeholders should be as small as possible to enable the development of a positive and effective destination image (Ryan and Aicken, 2010). Therefore, for the establishment of a relatively new tourism destination in this competitive tourism market, visitors' perceptions of the destination image must be studied and adjusted (Slak Valek and Williams, 2018). This is especially important when the destination must overcome certain negative elements of its image, as in the case of the selected destination, Linz (Austria), which has been struggling to escape from its associations with the Nazis and Hitler's perceptions of the world.

The aim of this study, therefore, is threefold. First, it uses the concept of 'Top of Mind' (Stepchenkova and Li, 2014), which has been adopted in tourism studies to test destination branding and image (Konecnik and Gartner, 2007; Kotsi and Valek, 2017). In the current study, the concept of Top of Mind is adopted to identify the spontaneous associations Linz evokes in visitors' minds and to highlight any differences that might exist between domestic and international visitors. Second, it aims to examine the gap between how domestic and international tourists perceive Linz's image (prior to visiting it and *in situ*) where the focus is on a set of image attributes linked to the DMO's attempt to reposition the destination in visitors' minds, such as its Nazi past and its cultural life (museums for modern art and festivals). Third, it seeks to highlight the differences in the image that occur because of visitors' actual experience to assess the success or failure of Linz's image repositioning. The chapter continues with an overview of the existing literature on destination image and its complex formation process.

9.2 Theoretical Framework

9.2.1 Definitions of destination image

The concept of image has been analysed from a variety of perspectives (Rodrigues *et al.*, 2011) and used in various disciplines, such as philosophy (Rodrigues *et al.*, 2011), psychology (Firestone and Scholl, 2015; Veldhuis *et al.*, 2020), marketing or branding (Arai *et al.*, 2014; Grohs and Reisinger, 2014; Chang *et al.*, 2019), and geography (Jenkins, 1999). Image in the field of tourism has spawned a diversity of definitions and conceptualizations, which could suggest that there is either a significant level of uncertainty as to what constitutes destination image and how it is formed, or that tourism destination image is a 'multi-dimensional and complex' (Gallarza *et al.*, 2002, p. 56) construct that can be embraced by all these definitions. One of the very first definitions of destination image was proposed by Hunt (1975) who described it as perceptions held by potential visitors about a destination. Later on, Crompton (1979) defined destination image as the sum of the ideas, beliefs and impressions a person has of a destination, which is one of the most prominent definitions of destination image to date and is the one adopted for the purposes of this study.

9.2.2 Destination image components and determinants

Although the term 'destination image' has been broadly used in tourism studies since the 1970s, its conceptualization is still considered an issue for a number of studies (Fakeye and Crompton, 1991; Gartner, 1993; Kim and Richardson, 2003; Rodrigues *et al.*, 2011; Zhang *et al.*, 2014; Kislali *et al.*, 2019). Boulding (1956), in his work *The Image: Knowledge in Life and Society*, in a dialogue with himself, proposes a theory of human behaviour from a psychological perspective based on perceptions of the world. He suggests that our knowledge of the world mirrors our image of the world since knowledge has an implication of validity and truth; hence what we believe is true is subjective and based on our own knowledge. Consequently, our actions depend on the image we have of the world and occur because of all our past experiences. He also postulates that people's subjective knowledge consists not only of images of 'fact' but also of images of 'value'. In other words, there is a difference between the image we hold of physical objects and our valuations of them, which is the way we evaluate the different parts of our image of the world. Boulding, therefore, was among the first who recognized the existence of cognitive (knowledge) and affective dimensions (emotions) of images.

In tourism literature, destination image is generally acknowledged to be the end-product of the fusion between cognitive and affective image elements. Knowledge/beliefs about a destination or even memories, evaluations and interpretations of a destination, represent the cognitive image components (Gartner, 1993; Baloglu, 1999; Pike and Ryan, 2004; Tasci *et al.*, 2007; Chen and Phou, 2013; Hallmann *et al.*, 2015). Affective components, in contrast, represent people's feelings or emotions towards a destination (Russell, 1980; Hanyu, 1993; Walmsley and Young, 1998; Yacout and Hefny, 2016). The conative component refers to the decision stage and is linked to future or actual behaviour or future intention towards a destination (Gartner, 1993; Baloglu and McCleary, 1999; Tasci *et al.*, 2007; Sahin and Baloglu, 2011; Hallmann *et al.*, 2015). The cognitive element is seen as more descriptive and measurable than the affective one (Walmsley and Young, 1998; Xie and Lee, 2013) and encompasses destination attractions such as cultural activities, traditions, landmarks, etc. (Beerli and Martin, 2004; Kim *et al.*, 2013; Eusebio and Vieira, 2013; Brito and Pratas, 2015; Stylos *et al.*, 2017; Iordanova and

Stylidis, 2017; Stylidis *et al.*, 2017). Previous research, in contrast, presented affective image components using four affective image attributes (distressing–relaxing, unpleasant–pleasant, boring–exciting, and sleepy–lively) on a semantic differential scale (Baloglu and McCleary, 1999; Martin and Bosque, 2008; Wang and Hsu, 2010) or Russell (1980) spatial model where affective evaluations are situated on four distinct continuums – arousal–sleepiness, misery–pleasure, distress–contentment, excitement–depression. The amalgamation of cognitive and affective components is seen as forming the overall image of the destination, which can be either positive or negative (e.g. Beerli and Martin, 2004; Li *et al.*, 2009).

The literature broadly acknowledges the impact of information sources, socio-demographic characteristics (age, education, gender, nationality/geographical location), psychological characteristics and socioeconomic status on the process of destination image formation (see Baloglu and McCleary, 1999; Baloglu, 2001; Beerli and Martin, 2004; Yu and Ko, 2012; Iordanova, 2015; Iordanova and Stylidis, 2017; Stylidis *et al.*, 2017; Andersen *et al.*, 2018). For example, Beerli and Martin (2004), Bonn *et al.* (2005) and Hsu *et al.* (2004) confirmed that the distance from a destination significantly affects its attribute-based and affective-based components of image. Respondents living far away from a destination were found to lack a vivid image of it (Reilly, 1990), but it is more favourable than the image that people who live close by might have (Crompton, 1979). The chapter continues with an overview of destination branding, which is deeply intertwined with destination image.

9.2.3 Destination branding

Zenker and Braun (2010) define place brand as 'a network of associations in the consumer's mind based on the visual, verbal, and behavioural expression of a place, which is embodied through the aims, communication, values, and the general culture of the place's stakeholders and the overall place design' (p. 5). This reflects broadly the Ritchie and Ritchie (1998) statement that destination brand is a name, symbol, logo, etc. that represents and differentiates the destination from its competitors.

Numerous destinations globally have attempted to rebrand themselves to refute negative place images (Bennett and Savani, 2003; Martınez, 2007; Gotham, 2007; Miller *et al.*, 2017). For example, Bennett and Savani (2003) reported on Amsterdam's 'City on the Water' rebranding campaign aimed at escaping its image of sexual liberalism and positioning it as a genteel tourist destination.

Difficult heritage, which is often seen to be human-made, is the main reason for some destinations to have an unattractive brand identity (Amujo and Otubanjo, 2012). Muhwezi *et al.* (2016) also suggest that places with difficult heritage should have a memorial emphasis rather than a commercial one when developing their positioning strategies aiming at changing deeply rooted prejudices and stereotypes in the minds of potential tourists by focusing on the commemorative, symbolic, functional, unique and distinctive attributes of the place.

Macdonald (2009) on Nazi Party rally grounds in Nuremberg, Germany, is one of the very few studies on destinations bearing the burden of its Nazi past, like the selected case study for this research. In her work, Macdonald reflects on Nuremberg's struggles to decide on the destiny of 'the largest single complex of monumental buildings ever constructed in National Socialist Germany' (p. 28). Through the years, local authorities' attitudes towards it went through 'silence' to suggestions to use it for motor racing, rock concerts or even transform it into a shopping centre, to finally deciding to incorporate it into its self-image and into projected image as there was 'widespread political consensus that Nuremberg's image would be harmed more by appearing to be not acknowledging its terrible past' (p.188). From the 1990s onwards, the city has attempted to promote itself as a major centre of commerce and culture and position itself as a City of Peace and Human Rights, while still dealing with its difficult heritage.

9.3 Case Study: Linz, Austria

The destination investigated in this research is Linz, Austria's third largest city, situated astride the Danube river and with a population of 205,000 citizens (AdminStat, 2019). Until the time of the First Republic, Linz was associated with provincial culture. During the Nazi period, however, Linz was transformed from a small town into an industrial city with a potential to become a cultural metropolis on the Danube, and in spite of the fact that Adolf Hitler was born in the outlying village of Braunau and only grew up in Linz, Linz became 'Hitler's town' as Salzburg is Mozart's. After 1945, the main concern of Linz's authorities was to distance themselves from Nazi culture and Hitler, while highlighting traditionally humanist cultural values (Linz Cultural Development Plan, 2000). Traces of the Nazi past, though, are still part of everyday life in Linz – in the appearance of the so-called Hitlerbauten (cheap homes introduced by Hitler for industrial workers), the industrial facilities of VOEST (leading European processing group with its own steel-making facilities) founded as the Hermann Göring Werke during World War II, and also in the materials used for the buildings' construction that raise an embarrassing point: Mauthausen granite was paid for with the lives of prisoners from the nearby concentration camp (Linz European Capital of Culture, 2009, cited in Iordanova-Krasteva *et al.*, 2010).

Linz's nomination for the title of European Capital of Culture for 2009 subtly signalled that the town no longer intends to be associated with its Nazi past and put the focus on its attempts since 1985 to reposition itself as a high-tech cultural city with a newly built museum for modern art and major hallmark events such as the Bruckner Festival, the International Street Artist Festival and the Ars Electronica Festival (Iordanova-Krasteva *et al.*, 2010). Nonetheless, the evaluation panel recommended that recent history should find a prominent place in the event which was in a sharp contrast with Linz's authorities' main concerns after 1945. Nevertheless, Linz's destination marketers embraced its dark history and sought to benefit from the association with Adolf Hitler, one of the last 'celebrities' to find a place in a destination promotion campaign (Pierce, 2009, cited in Iordanova-Krasteva *et al.*, 2010). In Linz, the local DMO has been striving to reposition the image of the town for many years. One of the first events of the European Capital of Culture programme, therefore, was an exhibition called 'The Fuhrer's Capital of Culture' and was part of a strategy aiming at overcoming Linz's dark history and audiences' prejudices. Ulrich Fuchs, the deputy manager of Linz09, said, with regard to Hitler's heritage, that 'whenever you come to Linz…you will find something related to this topic. We are not sweeping Hitler under the carpet' (Linz09 GmbH, 2010).

Developing this line of thought, Martin Heller, the artistic director of Linz09 (Linz09 GmbH, 2010), stated that 'we want to reflect back and show how cultural and political ambitions went together in the Nazi time…Talking about culture always means talking about politics…the only way of dealing with Hitler is to be completely honest.'

Moreover, the director of the Upper Austrian State Museums, Peter Assmann (Pierce, 2009), recognized that an exhibition about Linz's Hitler past might be going too far, because Hitler's legacy is still a very difficult and sensitive topic, but he defended the exhibition by arguing that, 'I don't see any glorification of Hitler in the exhibition. Hitler is fact, so we just face this fact and we face it with many arguments, with a lot of information about that time.'

These statements signal Linz's DMO readiness to embrace its burdensome past and incorporate it into their marketing campaigns, focusing on historical facts and lessons learnt from the past.

9.4 Methodology

Destination image is frequently assessed in the existing literature by using already-tested lists of image attributes or characteristics, which could lead to the omission of a destination's unique features and characteristics, such as, in the case of Linz, Hitler and the Nazi past. This study, therefore, uses a quasi mixed-method approach including both qualitative and quantitative techniques of data collection to capture all aspects of Linz's image as suggested by Echtner and Ritchie (1991) and Jenkins (1999). In this approach, unstructured techniques are commonly used to elicit the relevant destination image attributes, with researchers then using these attributes in subsequent analysis to construct surveys to investigate tourist images (O'Leary and Deegan, 2005). Existing studies used attributes associated with physical dimensions such as accommodation facilities, infrastructure, transportation, tourist facilities, climate/weather, scenery, natural environment, attractions and price (Echtner and Ritchie, 1991; Eusebio and Vieira, 2013; Pan *et al.*, 2014;

Brito and Pratas, 2015; Iordanova and Stylidis, 2017) to assess destination image.

The (first) qualitative data-collection stage took place online (e.g. online travel forums) and the purpose of this stage was mainly exploratory, which is reflected in the sample size, and aimed at revealing people's spontaneous associations with Linz as a tourist destination. These were then used to inform the second, quantitative stage of research. The key findings from the first data-collection stage and a subsequent review of the literature on destination image (e.g. Echtner and Ritchie, 1991; Baloglu and McCleary, 1999; Beerli and Martin, 2004; Prayag, 2009) were used to design a questionnaire to be used for collecting data from tourists (domestic and international, aged 18 or older) visiting Linz. Linz's image was assessed by asking respondents to evaluate on a Likert scale from 'Strongly Agree' to 'Strongly Disagree' its cultural life and Nazi heritage (steel/heavy industry and Hitler), since these two categories reflect the DMO's attempts to reposition Linz's image in visitors' minds. A non-probability sampling method was used due to the lack of accurate data on the size of the tourist population; nonetheless, the data collection took place at various locations in Linz and during different days/times of the week and at various locations to ensure the study's representativeness and reliability. Content analysis was used to analyse the answers to the open-ended questions and paired-sample t-tests were applied to gain insights into Linz's image change resulting from respondents' actual experience. The possible relationship between nationality (before and during visiting Linz) and Linz's cognitive image components was analysed using independent t-tests.

9.5 Results

Out of the 150 invited respondents, 88 agreed to participate and answer the questions included in the qualitative (first) phase. After discarding 14 incomplete responses, the final sample consisted of 74 usable responses that were content-analysed and the key findings were used to inform the questionnaire used in the second phase. Most of the respondents (74%) were from Great Britain, Germany, Switzerland, Portugal, Italy, Cyprus, France, Poland, Bulgaria and the USA, reflecting to a large extent the profile of international tourists in Linz. About half of the respondents were female and half were male.

The first question focused on respondents' spontaneous associations with Linz and aimed to capture the overall image of Linz as a tourist destination, and was answered by the majority (59) of the respondents. Only two Austrians (11%) left the question unanswered, compared to 13 international respondents (23%). Most of the respondents mentioned that Linz is in Austria and near the River Danube, showing that Linz benefits from the image of Austria as being a popular and favourite destination on the River Danube. However, the collected responses also indicate that it suffers in the shadow of Vienna and Salzburg and attempts to escape from

Austria's clichéd identity represented by those cities on the strength of its own merits (Linz09 GmbH, 2010). For example, one international respondent wrote: 'I associate Linz mainly with music and with the New Year concert of the Vienna Philharmonic Orchestra'. Traditional cuisine and music, as part of Linz's culture, found their place in the answers as well. However, this coin has a reverse side, as respondents said that Linz is just a small, old-fashioned Austrian town.

Indications that Linz is still paving the way from an industrial to a high-tech cultural city could be identified as a theme in some of the answers: 'unfortunately my first association with Linz is with the steel industry in the town'. It is of particular interest that the steel industry was mentioned only by Austrians, which indicates that nationality could impact, to some extent, people's perceptions of destinations, as stated in previous research (Beerli and Martin, 2004; Hsu *et al.*, 2004; Bonn *et al.*, 2005). Two domestic and two international respondents mentioned Hitler and Linz's dark Nazi past as their first associations with the place – a woman from France, for example, wrote: 'The first word that comes to my mind is Hitler'; and another male Austrian respondent stated: 'I associate this town mainly with Hitler and his "view" of the world...'.

Another question aimed at eliciting knowledge about some of Linz's attractions. Twenty-seven of the international respondents said that they do not have any knowledge about Linz. For others, the ancient origins of Linz, Hitler and its culture are interlinked: '...originated in the place of the ancient Rome town Lencia, it should become Hitler's capital city' and their knowledge about Linz is: 'Austria, Hitler and rich in cultural events'. The 21 international respondents mentioned different cultural events in Linz and museums. The number of respondents who mentioned Hitler increased to seven (only one was Austrian), thus assuming that knowledge of Linz's association with Hitler is not necessarily a negative factor in the formation of the overall image of the town, as feared by the local DMO.

A reliability analysis (Cronbach's alpha) was performed for the cognitive image scale, and the Cronbach's alpha value was higher than 0.70, which showed that the questionnaire achieved high internal consistency. The whole sample consisted of 400 respondents, 188 of which were Austrians (47% of the total sample) and 212 were internationals (53% of the total sample).

Respondents' spontaneous associations (Top of Mind concept) with Linz were evaluated (see Table 9.1) using an open-ended question where respondents were required to only write down keywords that come to their minds when hearing the word Linz. This question had two purposes: (i) to gain insights into their spontaneous image of Linz; and (ii) to assess Linz's branding richness and intensity by counting the number of spontaneous associations respondents have with it. This question was answered by all the respondents and the number of keywords varied between one to nine, with four keywords being the average number. From

Table 9.1. Respondents' profile.

	Sample $n = 400$
Gender	
Female	54%
Male	46%
Age	
18–25	8.5%
26–35	20.8%
36–45	23.5%
46–55	29.5%
56+	17.8%
Employment	
Full-time	57.5%
Part-time	16.3%
Student	7.3%
Retired	15.5%
Other	3.5%
Education	
Primary	7.3%
Secondary	35.5%
Tertiary	57.3%

the respondents, 35% wrote three keywords, followed by 29% with four keywords and 20% with five keywords. The nationality of the respondents appears to be in a relationship with the number of given answers since the number of internationals outweighed the number of Austrians within the group of respondents, with one to five associations, which represents nearly 80% of the whole sample. For example, 54% of the participants who had three spontaneous associations were internationals and 46% were domestic visitors, whereas from the group of respondents with five spontaneous associations, 67% were international respondents and only 33% were Austrians.

A content analysis was performed to analyse the spontaneous associations with Linz and the most popular category (235 responses, 55% internationals and 45% domestic visitors) was Museums and Galleries (including Linz's iconic Lentos and Ars Electronica museums). Among the most popular spontaneous associations with Linz, where 54% of the 56 responses were given by international visitors, were Linz's history heritage and its destiny to be Hitler's birthplace. One Japanese female respondent expressed, emotively: 'Hitler, concentration camps, grief and pain for Hitler's victims'. Similar statements came from several other respondents (mainly Austrians) who said that Mauthausen's concentration camp was one of their associations with Linz, or Hitler himself, since a few respondents called Linz 'Hitler's town'. Other fragments of Linz's history could

Table 9.2. Paired samples statistics.

		Mean	N	Std. Deviation
Pair 1	Bruckner Festival before	2.5550	400	1.67601
	Bruckner Festival after	2.8075	400	1.62390
Pair 2	Int. Street Artist Festival before	2.9300	400	1.82440
	Int. Street Artist Festival after	3.1950	400	1.68495
Pair 3	Cultural heritage before	1.5975	400	0.76613
	Cultural heritage after	1.2150	400	0.50439
Pair 4	Live music before	2.7475	400	1.64484
	Live music after	2.1750	400	1.48151
Pair 5	Ars Electronica Centre before	2.2675	400	1.30007
	Ars Electronica Centre after	1.4450	400	0.70585
Pair 6	Lentos before	2.2550	400	1.34518
	Lentos after	1.5475	400	0.70284
Pair 7	Modern Art before	2.7125	400	1.66637
	Modern Art after	1.6900	400	1.03275
Pair 8	Steel industry before	2.8475	400	1.42981
	Steel industry after	3.3375	400	1.15843
Pair 9	Heavy industry before	2.9150	400	1.44325
	Heavy industry after	3.4200	400	1.15626
Pair 10	Hitler before	2.2950	400	1.19228
	Hitler after	2.5925	400	1.17895

Mean scores are on a scale ranging from 1 (Strongly agree) to 5 (Strongly disagree).

also be identified in the following quotes despite the DMO's attempts to refute its difficult heritage: 'the old City Hall with the balcony where Hitler proclaimed the Greater German Reich'; 'Hitler's birthplace and its ambitions to expand the town further...'.

Another set of questions was designed to measure the level of agreement or disagreement with elicited responses in stage 1's cognitive image elements of Linz before and during respondents' actual experience in Linz, where a Likert scale was used (Strongly agree (1); Agree (2); Neutral (3); Disagree (4); Strongly Disagree (5)). A paired-sample t-test was applied to gain insights into Linz's image change resulting from respondents' actual experience. The results unveiled that significant changes in respondents' associations with Linz have occurred (Tables 9.2 and 9.3). For example, the mean values of Cultural Heritage before (M before = 1.59) and after (M after = 1.21) differed significantly. In a similar way, Linz's famous museums of modern art – Lentos and Ars Electronica Centre – also showed significant differences in respondents' opinions before and after their visit to Linz. For Ars Electronica Centre, the results were the following: on average, participants rated it significantly higher after their actual experience in Linz (M after = 1.54) than before (M before = 2.25), whereas Lentos Museum had

Table 9.3. Paired samples correlations.

Paired samples correlations					Paired samples test			
					Paired differences			
		N	Correlations	Sig.	Mean	Std. Deviation	t	Sig (2-tailed)
Pair 1	Bruckner Festival before & after	400	0.762	0	−0.2525	1.13885	−4.434	0
Pair 2	Int. Street Artist Festival before & after	400	0.748	0	−0.265	1.25248	−4.232	0
Pair 3	Cultural heritage before & after	400	0.367	0	0.3825	0.74672	10.245	0
Pair 4	Live music before & after	400	0.424	0	0.5725	1.6828	6.804	0
Pair 5	Ars Electronica Centre before & after	400	0.435	0	0.8225	1.17876	13.955	0
Pair 6	Lentos before & after	400	0.462	0	0.7075	1.19604	11.831	0
Pair 7	Modern Art before & after	400	0.474	0	1.0225	1.48746	13.748	0
Pair 8	Steel industry before & after	400	0.337	0	−0.49	1.50684	−6.504	0
Pair 9	Heavy industry before & after	400	0.358	0	−0.505	1.49182	−6.77	0
Pair 10	Hitler before & after	400	0.414	0	−0.2975	1.28379	−4.635	0

Mean scores are on a scale ranging from 1 (Strongly agree) to 5 (Strongly disagree)
*Significant at the level 0.05

the following significantly different values: M before = 2.25 compared to M after = 1.54. The museum for modern art changed with 1.02 on the scale. The average level of agreement for the Bruckner Festival (M before = 2.55, M after = 2.80) and the International Street Artist Festival (M before = 2.93, M after = 3.00) changed slightly, but still statistically significantly 'for worse' when comparing the pre- and post-travel results. All cognitive elements aligned with Linz's dark past showed significant decrease in the level of respondents' associations, demonstrating a positive change in Linz's image altogether. For example, the steel industry's mean value before visiting Linz was 2.84 and changed to 3.33, whereas the heavy industry's mean value before visiting Linz was 2.91 and decreased to 3.42. The association with Hitler was rated averagely at 2.29 before respondents' actual experience in Linz and went down to 2.59.

Table 9.4. Mean, SD and independent t-test prior to visiting Linz.

	Nationality/size	Mean	Std. Deviation		Levene's Test for Equality of Variances		t-test for Equality of Means	
					F	Sig.	t	Sig. (2-tailed)
Bruckner Festival before	Austrians 188	2.4096	1.6731	Equal variances assumed	0.114	0.735	–1.638	0.102
	Internationals 212	2.684	1.67194	Equal variances not assumed			–1.638	0.102
Int. Street Artist Festival before	Austrians 188	2.8511	1.87558	Equal variances assumed	1.218	0.271	–0.815	0.416
	Internationals 212	3	1.77929	Equal variances not assumed			–0.812	0.417
Cultural Heritage before	Austrians 188	1.6862	0.8729	Equal variances assumed	3.112	0.078	2.19	0.029
	Internationals 212	1.5189	0.64919	Equal variances not assumed			2.153	0.032
Live music before	Austrians 188	2.5957	1.73556	Equal variances assumed	6.597	0.011	–1.742	0.082
	Internationals 212	2.8821	1.55172	Equal variances not assumed			–1.73	0.084
Ars Electronica Centre before	Austrians 188	2.3777	1.27541	Equal variances assumed	0	0.988	1.599	0.111
	Internationals 212	2.1698	1.31684	Equal variances not assumed			1.602	0.11
Lentos before	Austrians 188	2.3936	1.43462	Equal variances assumed	6.328	0.012	1.948	0.052
	Internationals 212	2.1321	1.25116	Equal variances not assumed			1.932	0.054

Continued

Table 9.4. Continued.

	Nationality/size	Mean	Std. Deviation		Levene's Test for Equality of Variances		t-test for Equality of Means	
					F	Sig.	t	Sig. (2-tailed)
Modern Art before	Austrians 188	2.7979	1.731	Equal variances assumed	0.899	0.344	0.965	0.335
	Internationals 212	2.6368	1.6072	Equal variances not assumed			0.961	0.337
Steel industry before	Austrians 188	2.3032	1.10835	Equal variances assumed	36.728	0	-7.672	0
	Internationals 212	3.3302	1.50966	Equal variances not assumed			-7.812	0
Heavy industry before	Austrians 188	2.3404	1.10471	Equal variances assumed	38.038	0	-8.079	0
	Internationals 212	3.4245	1.51734	Equal variances not assumed			-8.23	0
Hitler before	Austrians 188	2.117	1.0275	Equal variances assumed	8.893	0.003	-2.836	0.005
	Internationals 212	2.4528	1.30353	Equal variances not assumed			-2.876	0.004

Mean scores are on a scale ranging from 1 (Strongly agree) to 5 (Strongly disagree).
*Significant at the level 0.05

The possible relationship between nationality (before and during visiting Linz) and Linz's cognitive image components were analysed using independent t-tests. The results indicated that prior to visiting Linz there is a statistically significant correlation between nationality and four out of the ten cognitive image elements, namely Cultural Heritage, Steel Industry, Heavy Industry and Hitler (Table 9.4). The mean for Cultural Heritage was nearly the same for Austrians (M = 1.68) and internationals (1.51); whereas for steel industry (Austrians M = 2.33, internationals M = 3.33), heavy industry (Austrians M = 2.34; internationals M = 3.42) and Hitler (Austrians M = 2.11; internationals M = 2.45) the differences in the means were more substantial, which indicates that prior to visiting Linz, the Austrian respondents were more likely to associate Linz with its negative side than were the internationals.

With respect to Linz's *in situ* image, all cognitive image elements that were previously found to be in a significant relationship with nationality, apart from Hitler (domestic M = 2.58; internationals M = 2.59), were found to still be significantly influenced by nationality. Moreover, significant differences were found in the answers given by the international and the domestic visitors in relation to Modern Art and Lentos (Table 9.5).

Cultural heritage results showed that internationals evaluate it more strongly (M = 1.12) and statistically significantly than Austrians (M = 1.31). Internationals associate Linz with Lentos almost to the same extent as Austrians (M = 1.47 compared to M = 1.62), albeit the difference is significantly different. This finding is mirrored by the fact that Internationals also accessed Linz's modern art significantly more than Austrians (M = 1.56 compared to M = 1.82). Results also indicated that Austrians are significantly more likely to associate Linz with its steel industry and heavy industry than internationals were.

9.6 Discussion and Conclusion

Positioning or repositioning a destination on the already overcrowded market is a challenging task in normal circumstances but paired with difficult heritage it could be a major obstacle for any DMO. The current study tentatively suggests that nationality influences the richness and intensity of respondents' associations with Linz where the internationals were more likely to have a higher number of spontaneous associations (keywords) than Austrians, which is in contrast with that of Reilly (1990) reporting that respondents living far away from a destination were found to lack a vivid image of it. Moreover, the Nazi past appears to be still deeply rooted in Linz's brand despite the DMO's continuous attempts to reposition the destination's image by consciously or subconsciously trying to refute its difficult heritage and focusing its marketing efforts on promoting Linz's vivid cultural life.

It was also revealed that actual experience at the destination positively modifies and influences its image, as suggested in the literature (Baloglu, 2001). Nevertheless, the modifications that occurred were more substantial

Table 9.5. Mean, SD and independent t-test during the stay in Linz.

Nationality		Mean	Std. Deviation		Levene's Test for Equality of Variances		t-test for Equality of Means	
					F	Sig.	t	Sig. (2-tailed)
Bruckner Festival after	Austrians	2.7447	1.70207	Equal variances assumed	3.275	0.071	–0.728	0.467
	Internationals	2.8632	1.55322	Equal variances not assumed			–0.724	0.469
Int. Street Artist Festival after	Austrians	3.1436	1.79292	Equal variances assumed	6.233	0.013	–0.574	0.566
	Internationals	3.2406	1.58597	Equal variances not assumed			–0.57	0.569
Cultural heritage after	Austrians	1.3191	0.63267	Equal variances assumed	56.856	0	3.96	0
	Internationals	1.1226	0.3288	Equal variances not assumed			3.825	0
Live music after	Austrians	2.2979	1.66004	Equal variances assumed	10.379	0.001	1.565	0.118
	Internationals	2.066	1.29724	Equal variances not assumed			1.542	0.124
Ars Electronica after	Austrians	1.5	0.68988	Equal variances assumed	1.33	0.249	1.47	0.142
	Internationals	1.3962	0.71781	Equal variances not assumed			1.473	0.141
Lentos after	Austrians	1.6277	0.73832	Equal variances assumed	2.506	0.114	2.158	0.032
	Internationals	1.4764	0.66348	Equal variances not assumed			2.144	0.033
Modern Art after	Austrians	1.8298	1.18492	Equal variances assumed	6.105	0.014	2.567	0.011

Continued

Table 9.5. Continued.

Nationality		Mean	Std. Deviation		Levene's Test for Equality of Variances		t-test for Equality of Means	
					F	Sig.	t	Sig. (2-tailed)
	Internationals	1.566	0.86005	Equal variances not assumed			2.52	0.012
Steel industry after	Austrians	3.1809	1.13724	Equal variances assumed	1.044	0.308	-2.565	0.011
	Internationals	3.4764	1.16196	Equal variances not assumed			-2.568	0.011
Heavy industry after	Austrians	3.2553	1.14636	Equal variances assumed	0.178	0.673	-2.704	0.007
	Internationals	3.566	1.148	Equal variances not assumed			-2.704	0.007
Hitler after	Austrians	2.5851	1.13674	Equal variances assumed	0.052	0.819	-0.118	0.906
	Internationals	2.5991	1.2178	Equal variances not assumed			-0.118	0.906

Mean scores are on a scale ranging from 1 (Strongly agree) to 5 (Strongly disagree).
*Significant at the level 0.05

for the international visitors, which highlights the need for a differentiated tourism product and marketing campaign to adjust Linz's image for the domestic market and overcome any stigma associated with the town. Overall, Linz's image attributes, linked to its DMO's strategies to establish the town as a centre for modern art and contemporary culture, were rated higher after visiting it, which signals that the DMO's efforts to rebrand the destination are starting to be fruitful. This is supported by the fact that respondents' associations with Linz's difficult heritage after visiting it significantly decreased, which again demonstrates a positive change in its image. This illustrates that rectifying destination image is feasible in the long term if destinations with difficult heritage engage in an unprejudiced discourse and present a balanced, non-judgemental and facts-based account of past events.

The current study also verified the results of previous studies (Hunt, 1975; Fakeye and Crompton, 1991; MacKay and Fesenmaier, 2000; Hsu *et al.*, 2004; Beerli and Martin, 2004) that 'distance' (or 'nationality'), in general, has a direct effect over tourism destination images. Nevertheless, it was also shown that different stakeholders see the destination differently and targeting both domestic and international visitors using the same promotion strategy would not lead to the desired outcomes. This finding is adherent with Crompton (1979) that individuals who live away from a destination have a tendency to hold more positive images of it, since the Austrians were found to have a more negative image of Linz than the internationals visitor. For example, the main differences between the Austrians and the internationals, prior to visiting Linz, were found to be linked to Linz's difficult past (Hitler, heavy industry, steel industry, cultural heritage), indicating that the Austrians were more likely to associate Linz with its negative aspects than were the internationals. The internationals rated Linz's *a priori* cognitive image dimension 'Contemporary Culture' higher than the Austrians (not significantly, though), showing that the Austrians tend to underestimate Linz's place on the European cultural map of modern art. This finding supports Bonn *et al.*'s (2005) argument that destinations should not use the same destination image to attract visitors with different sociocultural backgrounds. Not surprisingly, Linz's *in situ* image showed similar results with some minor alterations. Austrians were still more likely to have slightly more negative associations with Linz than internationals whose associations were mostly linked to Linz's contemporary art image elements (Modern Art, Lentos). The reported findings call for a more targeted campaign on a local/domestic level aimed at removing prejudices and promoting Linz as a cultural destination for the domestic market. This could be achieved by using, at a national level, traditional promotional channels such as radio, television and tourism brochures, but also by involving local renowned Austrians to act as destination ambassadors.

In addition to these recommendations, the local DMO should engage in a dialogue with potential domestic tourists on various social media sites where the focus should be on rebranding the destination by promoting its lively cultural life and natural beauty. Moreover, special events of an educational nature concerning Linz's history could be introduced aimed at 'healing' Linz's negative image, which, as this study suggests, is already happening. Surprisingly, there were no longer any significant differences among respondents' associations with Hitler. This finding implies that Linz's DMO's decision to incorporate the Nazi past into various exhibitions and be open and honest about it (similarly Nuremberg; Macdonald, 2009), neutralized any negative connotations it might have had prior to this. Linz, as other destinations with difficult heritage, should, instead of ignoring it, embrace it in their brand strategies in a mature, facts-based manner and aim to change prejudices and stereotypes in the minds of potential visitors (Amujo and Otubanjo, 2012; Muhwezi *et al.*, 2016). DMOs facing such difficult issues could use them as an opportunity to develop tours aimed at revealing the past in an objective way, so as to establish the foundation for new beginnings.

This study is not free from limitations that could have an impact on its generalizability, but this does not invalidate its findings. First, the Nazi past is unique to Linz and it would be difficult to replicate it in other destinations; however, other places burdened with Nazi history could learn from Linz's attempts to transform itself to towards a place associated with contemporary art and culture. Second, the international respondents represented mainly European countries, which may have distorted the results, and a more heterogeneous sample might be more beneficial. Third, a topic as sensitive as the Nazis calls for a qualitative discourse where respondents' emotions and feelings are investigated more in depth, which could lead to a better understanding of the differences that occurred in the way international and domestic visitors perceive Linz. The current study shows that repositioning a destination with a burdensome past is still feasible, though challenging, with well-thought-through, targeted marketing campaigns, and other destinations could learn from these attempts to reposition Linz as a cultural, modern tourism destination. Future research could expand our understanding of the integration of difficult heritage in destination image and branding strategies by replicating this study in a different context using different research approaches and sampling techniques to validate the current findings and propose further ways to reposition a destination with a burdensome past.

References

AdminStat (2019) Maps, analysis and statistics about the resident population. Available at: https://ugeo.urbistat.com/AdminStat/en/at/demografia/popolazione/linz-stadt-/20132882/4 (accessed 20 December 2020).

Agapito, A., Mendes, J. and Valle, P. (2010) Destination image perspectives of tourists versus residents. *European Journal of Tourism, Hospitality and Recreation* 1, 90–109.

Ahn, T., Ekinci, Y. and Li, G. (2013) Self-congruence, functional congruence, and destination choice. *Journal of Business Research* 66, 719–723.

Alcocer, N.H. and Ruiz, V.Z. (2020) The role of destination image in tourist satisfaction: the case of a heritage site. *Economic Research-Ekonomska Istraživanja* 33, 2444–2461.

Amujo, O. and Otubanjo, O. (2012) Leveraging rebranding of 'unattractive' nation brands to stimulate post-disaster tourism. *Tourist Studies* 12, 89–105.

Andersen, O., Oian, H., Aas, O. and Tangeland, T. (2018) Affective and cognitive dimensions of ski destination images: the case of Norway and the Lillehammer region. *Scandinavian Journal of Hospitality and Tourism* 28, 113–131.

Arai, A., Ko, Y. and Ross, S. (2014) Branding athletes: exploration and conceptualization of athlete brand image. *Sport Management Review* 17, 97–106.

Baloglu, S. (1999) A path analytic model of visitation intention involving information sources, socio-psychological motivations, and destination image. *Journal of Travel and Tourism Marketing* 8, 81–91.

Baloglu, S. (2001) Image variations of Turkey by familiarity index: informational and experiential dimensions. *Tourism Management* 22, 127–133.

Baloglu, S. and McCleary, K. (1999) A model of destination image formation. *Annals of Tourism Research* 26, 868–897.

Beerli, A. and Martin, J. (2004) Factors influencing destination image. *Annals of Tourism Research* 31, 657–681.

Bennett, R. and Savani, S. (2003) The rebranding of city places: an international comparative investigation. *International Public Management Review* 4, 70–87.

Bigné, J., Sanchez, M. and Sanchez, J. (2001) Tourism image, evaluation variables and after purchase behaviour: inter-relationships. *Tourism Management* 22, 607–616.

Bonn, M.A., Joseph, S.M. and Dai, M. (2005) International versus domestic visitors: an examination of destination image perceptions. *Journal of Travel Research* 43, 294–301.

Boulding, K.E. (1956) *The Image: Knowledge and Life in Society.* University of Michigan Press, Ann Arbor, Michigan.

Brito, P. and Pratas, J. (2015) Tourism brochures: linking message strategies, tactics and brand destination attributes. *Tourism Management* 48, 123–138.

Buhalis, D. (2000) Marketing the competitive destination of the future. *Tourism Management* 21, 97–116.

Carballo, M.M., Araña, J.E., León, C.J. and Moreno-Gil, S. (2015) Economic valuation of tourism destination image. *Tourism Economics* 21, 741–759.

Chang, Y., Li, Y., Yan, J. and Kumar, V. (2019) Getting more likes: the impact of narrative person and brand image on customer–brand interactions. *Journal of the Academy of Marketing Science* 47, 1027–1045.

Chen, C. and Phou, S. (2013) A closer look at destination: image, personality, relationship and loyalty. *Tourism Management* 36, 269–278.

Chen, C.F. and Tsai, D. (2007) How destination image and evaluative factors affect behavioural intentions? *Tourism Management* 28, 1115–1122.

Chen, P.-J. and Kerstetter, D.L. (1999) International students' image of rural Pennsylvania as a travel destination. *Journal of Travel Research* 37, 256–266.

Chi, C.G.Q. and Qu, H. (2008) Examining the structural relationships of destination image, tourist satisfaction and destination loyalty: an integrated approach. *Tourism Management* 29, 624–636.

Crompton, J.L. (1979) Motivations for pleasure vacation. *Annals of Tourism Research* October, 408–424.

Crouch, G. and Ritchie, J. (1999) Tourism, competitiveness, and social prosperity. *Journal of Business Research* 44, 137–152.

Dumont, E., Asensio, M. and Mortari, M. (2010) Image, construction and representation in tourism promotion and heritage management. In: Burns, P., Palmer, C. and Lester, J.A. (eds) *Tourism and Visual Culture. Volume 1: Theories and Concepts.* CAB International, Wallingford, UK, pp. 124–138.

Echtner, C. and Ritchie, J. (1991) The meaning and measurement of destination image. *The Journal of Tourism Studies* 2, 2–12.

Echtner, C. and Ritchie, J. (1993) The measurement of destination image: an empirical assessment. *Journal of Travel Research* 31, 3–13.

Eusebio, C. and Vieira, A. (2013) Destination attributes' evaluation, satisfaction and behavioural intentions: a structural modelling approach. *International Journal of Tourism Research* 15, 66–80.

Fakeye, P. and Crompton, J. (1991) Image differences between prospective, first-time, and repeat visitors to the Lower Rio Grande Valley. *Journal of Travel Research* 30, 10–16.

Fernando, I. and Long, W. (2012) New conceptual model on cluster competitiveness: a new paradigm for tourism? *International Journal of Business and Management* 7, 75–84.

Firestone, C. and Scholl, B. (2015) Cognition does not affect perception: evaluating the evidence. *Behavioral and Brain Sciences* 39, 1–77.

Gallarza, M., Saura, I. and Garcia, H. (2002) Destination image: towards a conceptual framework. *Annals of Tourism Research* 29, 56–78.

Gartner, W.C. (1993) Image formation process. *Journal of Travel and Tourism Marketing* 2, 191–215.

Gotham, K. (2007) Re)Branding the big easy: tourism rebuilding in post-Katrina New Orleans. *Urban Affairs Review* 42, 823–850.

Grohs, R. and Reisinger, H. (2014) Sponsorship effects on brand image: the role of exposure and activity involvement. *Journal of Business Research* 67, 1018–1025.

Hallmann, K., Zehrer, A. and Müller, S. (2015) Perceived destination image: an image model for a Winter Sports destination and its effect on intention to revisit. *Journal of Travel Research* 54, 94–106.

Hanyu, K. (1993) The affective meaning of Tokyo: verbal and nonverbal approaches. *Journal of Environmental Psychology* 13, 161–172.

Hsu, C.H.C., Wolfe, K. and Kang, S.K. (2004) Image assessment for a destination with limited comparative advantages. *Tourism Management* 25, 121–126.

Hunt, J. (1975) Image as a factor in tourism development. *Journal of Travel Research* 13, 1–7.

Iordanova, E. (2015) Unravelling the complexity of destination image formation: a conceptual framework. *European Journal of Tourism Research* 11, 35–56.

Iordanova, E. and Stylidis, D. (2017) International and domestic tourists' 'a priori' and 'in situ' image differences and the impact of direct destination experience on destination image: the case of Linz, Austria. *Current Issues in Tourism* 9, 0–24.

Iordanova-Krasteva, E., Wickens, E. and Bakir, A. (2010) The ambiguous image of Linz. *PASOS* 8, 67–77.

Jenkins, O. (1999) Understanding and measuring tourist destination images. *International Journal of Tourism Research* 1, 1–15.

Kim, H. and Richardson, S. (2003) Motion picture impacts on destination images. *Annals of Tourism Research* 30, 216–237.

Kim, S., Holland, S. and Han, H. (2013) A structural model for examining how destination image, perceived value, and service quality affect destination loyalty: a case study of Orlando. *International Journal of Tourism Research* 15, 313–328.

Kislali, H., Kavaratzis, M. and Saren, M. (2019) Destination image formation: towards a holistic approach. *International Journal of Tourism Research* 22, 266–276.

Kneesel, E., Baloglu, S. and Millar, M. (2010) Gaming destination images: implications for branding. *Journal of Travel Research* 49, 68–78.

Konecnik, M. (2002) The image as a possible source of competitive advantage of the destination – the case of Slovenia. *Tourism Review* 57, 6–12.

Konecnik, M. and Gartner, W.C. (2007) Customer-based brand equity for a destination. *Annals of Tourism Research* 34, 400–421.

Kotsi, F. and Valek, N.S. (2017) Flying with Nicole Kidman or Jennifer Aniston? Brand funnel stages' influence on brand personality. *Journal of Travel and Tourism Marketing*, 1–12.

Kozak, M. (2001) Repeaters' behaviour at two distinct destinations. *Annals of Tourism Research* 28, 784–807.

Lee, B., Lee, C.K. and Lee, J. (2014) Dynamic nature of destination image and influence of tourist overall satisfaction on image modification. *Journal of Travel Research* 53, 239–251.

Li, X., Pan, B., Zhang, L. and Smith, W. (2009) The effect of online information search on image development: insights from a mixed-methods study. *Journal of Travel Research* 48, 45–57.

Linz09 GmbH (2010) Final Report Linz09. Linz 2009 European Capital of Culture. Available at: http://www.linz09.at/sixcms/media.php/4974/Final%20Report_en_TOTAL.680790.pdf (accessed 30 June 2021).

Linz Cultural Development Plan (2000) Resolution of the Linz City Council. Available at: http://www.linz.at/images/Cultural_Development_Plan_2000.pdf (accessed 30 June 2021).

Linz European Capital of Culture (2009) Mission statement. Available at: http://www.linz09.at/en/mission-statement.html (accessed 30 June 2021).

Macdonald, S. (2009) *Difficult Heritage: Negotiating the Nazi Past in Nuremberg and Beyond*. Routledge, London and New York.

MacKay, K.J. and Fesenmaier, D.R. (2000) An exploration of cross-cultural destination image assessment. *Journal of Travel Research* 38, 417–423.

Martin, H.S. and Bosque, I.A.R. (2008) Exploring the cognitive-affective nature of destination image and the role of psychological factors in its formation. *Tourism Management* 29, 263–277.

Martınez, J. (2007) Selling avant-garde: how Antwerp became a fashion capital (1990–2002). *Urban Studies* 44, 2449–2464.

Mayo, E. and Jarvis, L. (1981) *The Psychology of Leisure Travel*. CBI, Boston, Massachusetts.

Miller, D.S., Gonzalez, S. and Hutter, M. (2017) Phoenix tourism within dark tourism rebirth, rebuilding and rebranding of tourist destinations following disasters. *Worldwide Hospitality and Tourism Themes* 9, 196–215.

Muhwezi, D., Baum, T. and Nyakaana, J. (2016) Dealing with negative symbolism of destinations with difficult heritage: analysis of Uganda's image. *Journal of Hospitality Management and Tourism* 7, 33–42.

O'Leary, S. and Deegan, J. (2005) Ireland's image as a tourism destination in France: attribute importance and performance. *Journal of Travel Research* 43, 247–256.

Pan, S. (2011) The role of TV commercial visuals in forming memorable and impressive destination images. *Journal of Travel Research* 50, 171–185.

Pan, S., Lee, J. and Tsai, H. (2014) Travel photos: motivations, image dimensions, and affective qualities of places. *Tourism Management* 40, 59–69.

Pierce, A. (2009) Adolf Hitler at the centre of Austria's City of Culture campaign. Available at: http://www.telegraph.co.uk/news/worldnews/europe/austria/4076893/Adolf-Hitler-at-the-centre-of-Austrias-City-of-Culture-campaign.html (accessed 30 June 2021).

Pike, S. and Ryan, C. (2004) Destination positioning analysis through a comparison of cognitive, affective, and conative perceptions. *Journal of Travel Research* 42, 333–342.

Prayag, G. (2009) Tourists' evaluation of destination image, satisfaction and future behavioural intentions: the case of Mauritius. *Journal of Travel and Tourism Marketing* 26, 836–853.

Qu, H., Kim, L.H. and Im, H.H. (2011) A model of destination branding: integrating the concepts of the branding and destination image. *Tourism Management* 32, 465–476.

Reilly, M.D. (1990) Free elicitation of descriptive adjectives for tourism image assessment. *Journal of Travel Research* 28, 69–76.

Ritchie, J. and Crouch, G. (2010) A model of destination competitiveness/sustainability: Brazilian perspectives. *Revista de Administração Pública* 44, 1049–1066.

Ritchie, J. and Ritchie, R. (1998) The branding of tourism destinations – past achievements and future challenges. In: Keller, P. (ed.) *Destination Marketing – Reports of the 48th AIEST Congress*. Marrakech, pp. 89–116.

Rodrigues, A.I., Correia, A. and Kozak, M. (2011) A multidisciplinary approach on destination image construct. *Tourismos: An International Multidisciplinary Journal of Tourism* 6, 93–110.

Russell, J. (1980) A circumplex model of affect. *Journal of Personality and Social Psychology* 39, 1161–1178.

Ryan, C. and Aicken, M. (2010) The destination image gap – visitors' and residents' perceptions of place: evidence from Waiheke Island, New Zealand. *Current Issues in Tourism* 13, 541–561.

Sahin, S. and Baloglu, S. (2011) Brand personality and destination image of Istanbul. *Anatolia* 22, 69–88.

Slak Valek, N. and Williams, R. (2018) One place, two perspectives: destination image for tourists and nationals in Abu Dhabi. *Tourism Management Perspectives* 27, 152–161.

Stepchenkova, S. and Li, X.R. (2014) Destination image: do top-of-mind associations say it all? *Annals of Tourism Research* 45, 46–62.

Stylidis, D., Belhassen, Y. and Shani, A. (2015) Three tales of a city: stakeholders' images of Eilat as a tourist destination. *Journal of Travel Research* 54, 702–716.

Stylidis, D., Shani, A. and Belhassen, Y. (2017) Testing an integrated destination image model across residents and tourists. *Tourism Management* 58, 184–195.

Stylos, N., Bellou, V., Andronikidis, A. and Vassiliadis, C.A. (2017) Linking the dots among destination images, place attachment, and revisit intentions: a study among British and Russian tourists. *Tourism Management* 60, 15–29.

Tasci, A.D.A. and Kozak, M. (2006) Destination brands vs destination images: Do we know what we mean? *Journal of Vacation Marketing* 12, 299–317.

Tasci, A.D.A., Gartner, W.C. and Cavusgil, S. (2007) Conceptualization and operationalization of destination image. *Journal of Hospitality and Tourism Research* 31, 194–223.

Veldhuis, J., Alleva, J., Bij de Vaate, A., Keijer, M. and Konijn, E. (2020) Me, my selfie, and I: the relations between selfie behaviors, body image, self-objectification, and self-esteem in young women. *Psychology of Popular Media Culture* 9, 3.

Walmsley, D. and Young, M. (1998) Evaluative images and tourism: the use of personal constructs to describe the structure of destination images. *Journal of Travel Research* 36, 65–69.

Wang, C. and Hsu, M. (2010) The relationships of destination image, satisfaction, and behavioural intentions: an integrated model. *Journal of Travel and Tourism Marketing* 27, 829–843.

Xie, K. and Lee, J. (2013) Toward the perspective of cognitive destination image and destination personality: the case of Beijing. *Journal of Travel and Tourism Marketing* 30, 538–556.

Yacout, O. and Hefny, L. (2016) Use of Hofstede's cultural dimensions, demographics, and information sources as antecedents to cognitive and affective destination image for Egypt. *Journal of Vacation Marketing* 21, 37–52.

Yu, J. and Ko, T. (2012) A cross-cultural study of perceptions of medical tourism among Chinese, Japanese and Korean tourists in Korea. *Tourism Management* 33, 80–88.

Zenker, S. and Braun, E. (2010) Towards an integrated approach for place brand management. Paper presented at the 50th European Regional Science Association Congress, Jönköping, Sweden. Available at: http://www-sre.wu.ac.at/ersa/ersaconfs/ersa10/ERSA2010finalpaper181.pdf (accessed 5 November 2020).

Zhang, H., Fu, X., Cai, L. and Lu, L. (2014) Destination image and tourist loyalty: a meta-analysis. *Tourism Management* 40, 213–223.

10 Being Online Off the Beaten Track: Rhythms of Connectivity and User-generated Content in the Marketing of Greenland

Carina Ren[1]* and Elizabeth Cooper[2]

[1]*Aalborg University, Denmark*
[2]*Visit Greenland/Copenhagen Business School, Denmark*

10.1 Introduction

In recent years and in the light of climate changes, many destinations in the Arctic parts of Europe have been successful in capitalizing on an increased interest in polar tourism. Leading up to the Covid-19 crisis, visitor numbers remained modest in Greenland, a self-ruled Arctic nation under the Danish Commonwealth. However, as the nation's tourism rebuild coincides with the opening of two new international airports in 2024, visitor numbers are expected to rise higher than ever before. Successful marketing, combined with a rising global focus on the Arctic, has led to a growing volume of tourists and a corresponding rise in political and societal interest in tourism. Already, today, tourism is viewed as one of Greenland's three economic pillars, next to the dominant fishing industry and mining. Also, it is perceived by many as a promising lever for the Arctic nation's future economic development (Bjørst and Ren, 2015).

The tourism offerings in Greenland are predominantly adventure tourism, and, similar to other Arctic destinations, the experiential core is rested, as argued by Lee *et al.* (2017) in 'a context of geographical peripherality' (p.1). In other words, in being and feeling, as a tourism experience, truly off the beaten track. As an Arctic destination heavily affected by climate change, Greenland has witnessed increasing interest from a global public, something that is believed to have helped bring the destination closer in the mind of potential visitors. The destination has also received unexpected media attention and exposure as a result of growing geopolitical interest in

*Corresponding author: ren@hum.aau.dk

 Tourism Marketing in Western Europe (eds N. Boukas and D. Stylidis)
DOI: 10.1079/9781789248753.0010

the Arctic: for instance when US president, Donald Trump, proposed to buy the island in 2019. The national DMO, Visit Greenland, has been quick to make the most of this global access and the general rising interest on social media. In securing continued growth, efficient marketing of the destination has been identified by Visit Greenland as essential. Visit Greenland markets its products and experiences around the 'big Arctic five': dogsledding, whale-watching, Northern Lights, ice/snow, and the nation's pioneering people. In combination with sailing, hiking, fishing and hunting, as well as cultural offerings, such as kaffemik coffee parties in private homes and visits to handicraft workshops, these comprise the products offered to the nearly 100,000 people who visit Greenland annually (Statistics Greenland, 2019).

To better understand how Visit Greenland works in marketing Greenland as a tourist destination, the case of social media influencers is explored. To do so, a review of the social media marketing literature is provided which zooms in on social media marketing, tourism and social media influencers in tourism. Encapsulation and decapsulation is introduced as an analytical tool to explore the study's material and is presented in the following analysis. After a short introduction to our methodology and case, empirical material from interviews with professional social media influencers sponsored by the DMO to promote the destination is used to show how users negotiate online presence and social media content during their travels in Greenland. In conclusion, the trade-offs and balancing acts of being 'online' and 'off the beaten track' are discussed, seeing them as equally important but at times hard to reconcile.

10.2 Literature Review

10.2.1 Social media marketing – a very brief history

The sheer speed at which information technologies and social networking platforms have been developing over the last few decades has led to much existing literature being largely descriptive. Accounts of new developments and resulting changes in organizational and user behaviour dominate the literature – and this is somewhat unsurprising, given that new developments naturally invite description and definition. For this reason, we will only briefly define the relevant historical developments before moving on to a more detailed discussion of the literature which specifically shapes this research.

One of the key turns in information technology has been the evolution from Web 1.0 to Web 2.0, which saw the internet shift from a one-to-many communication platform to a many-to-many communication platform (Schegg *et al.*, 2008; Hays *et al.*, 2013; Minazzi, 2015). This has given users of the internet more power, as they are able to create and disseminate their own content (user-generated content) and to participate in the creation of information with their own thoughts and opinions (Buhalis and Law, 2008; Xiang and Gretzel, 2010; Hays *et al.*, 2013). Building on Web 2.0, social media platforms have allowed the sharing of user-generated content within participatory and fluid online communities (Kaplan and Haenlein, 2010; Hays *et al.*, 2013; Minazzi, 2015). These platforms have facilitated the evolution from traditional word-of-mouth marketing to eWOM (electronic word-of-mouth) or 'world-of-mouth' (Qualman, 2012; Seeler *et al.*, 2019). Because of this development, it is argued that customer-to-customer recommendations have gained importance and now extend beyond direct social circles to include relationships that only exist on a digital level (Lin, 2018; Seeler *et al.*, 2019).

Social media marketing sees marketers utilizing social media platforms in order to generate awareness of and stimulate conversation about their products (Parise *et al.*, 2008; Hays *et al.*, 2013). According to Statista Market Forecast (2019), brands worldwide spent nearly US$90 million on social media marketing in 2019 (Statista Market Forecast, 2019; De Jans *et al.*, 2020). A particularly popular social media marketing platform is Instagram (Phua *et al.*, 2017), which had more than one billion active monthly users in 2019 (Omnicore Agency, 2019; De Jans *et al.*, 2020). The multi-directional nature of social media marketing, as well as the growth of platforms that blend marketing with user-generated content, renders this landscape increasingly complex and full of tensions, which will be further explored below.

10.2.2 Social media marketing and tourism

Tourism marketing has evolved alongside, and intertwined with, social media marketing with an evolution from Travel 1.0 to Travel 2.0 in which, according to some scholars, travellers are increasingly being inspired by user-generated content and moving their purchasing habits online (Hays *et al.*, 2013; Minazzi, 2015; Seeler *et al.*, 2019). The significant influence of eWOM (Minazzi, 2015), the increasing value of online travel purchases (Xu and Pratt, 2018) and the relative low cost and wide reach of social media marketing (Hays *et al.*, 2013) has, in recent years, led many DMOs to develop marketing strategies that heavily feature social media (Schegg *et al.*, 2008; Hays *et al.*, 2013). The limited literature on DMOs' use of social media marketing tends to focus on DMOs' own social media presence through the analysis of DMO accounts (Hays *et al.*, 2013). However, the influence of DMOs on the social media presence of their destination in reality extends far beyond their own social media accounts and deserves in-depth analysis. One of the ways that DMOs' influence extends more subtly into the social media marketing mix is by using social media influencers.

10.2.3 Defining social media influencers: when influencer marketing meets tourism

The increasing use of social media marketing has led to the emergence of social media influencers (SMIs) (Seeler *et al.*, 2019). There is general agreement in the literature that SMIs are defined as independent third-party endorsers who influence audience attitudes via the use of social media (Freberg *et al.*, 2011; Dhanesh and Duthler, 2019; Magno and Cassia, 2018). They are often specialized in a particular field and have large and loyal follower bases who are influenced, often to a large extent, by their content (De Veirman *et al.*, 2017; Fäs and Zumstein, 2019; De Jans *et al.*, 2020). Some scholars have defined SMIs as 'everyday people' (Xu and Pratt, 2018, p. 958) using social media to document their 'daily lives' (De Jans *et al.*, 2020, p. 1), although this is where the definition becomes debatable and worthy of further investigation. In reality, nowadays, their social media presence is their full-time job for many SMIs, so their daily life is arguably shaped more by their social media presence than the other way round. Since it is also widely agreed that many SMIs nowadays can be considered minor celebrities (Kaplan and Haenlein, 2012; Xu and Pratt, 2018), it is also highly questionable to refer to them as 'everyday people'. It is perhaps more accurate to argue that SMIs fall somewhere in between a celebrity and an acquaintance in terms of their relationship to their followers (Chae *et al.*, 2017; Dhanesh and Duthler, 2019).

Influencer marketing is now widely used across different industries. According to Farivar *et al.* (2019), Instagram is, at the time of writing, the most used platform for influencer marketing. In influencer marketing, companies collaborate with SMIs, who promote the companies' products in return for sponsorship or payment (Farivar *et al.*, 2019, p. 1). Thus, the SMIs' unique relationship to their followers is leveraged for brand or product promotion (Seeler *et al.*, 2019). Numerous studies of influencer marketing have attempted to investigate the extent to which it is effective or actually influences purchase intentions (Seeler *et al.*, 2019; Cuevas *et al.*, 2020).

However, literature that specifically analyses the use of influencer marketing in tourism is scarce (Magno and Cassia, 2018; Seeler *et al.*, 2019; Stoldt *et al.*, 2019), and it is precisely when we apply existing understandings of SMIs to tourism that definitions no longer seem to fit and boundaries become blurred. SMIs who specialize in tourism do not tend to post their daily lives; on the contrary, their social media presence is primarily based around their travels, or on what is traditionally thought of – but increasingly contested – as the 'escape' from everyday life.

Tourism SMIs do not advertise brands in the traditional sense. eWOM, in the case of a destination, is arguably even more subtle as no brand names need be mentioned. Indeed, even the name of the DMO can be completely avoided, since the destination itself is the product. In other words, as long as the SMIs make clear where they are, all other signs of sponsorship or collaboration with a company can be hidden. This can make it difficult or even impossible for followers to determine whether content has been sponsored or whether this is simply the influencers posting about their holiday – thereby complicating the role of the tourism SMI even further. It is perhaps partly because of this ambiguity that influencer marketing in tourism remains relatively unexplored. However, there are still more complexities that arise when traditional tourism scholarship meets the more contemporary body of SMI scholarship. Some of these are explored in more detail below.

10.2.4 Encapsulation and decapsulation: then and now

As outlined in the introduction, this chapter aims to explore the 'live' element of influencer marketing, and the way in which this affects the task of promoting a country like Greenland, where an internet connection is not always available. Existing literature that specifically explores the nature of 'being live' as an SMI (and goes beyond descriptions of live-streaming services) is hard to find, but we can certainly shed light on this experience by looking back on older texts that discuss how holidays are experienced and subsequently retold.

Connectivity has been discussed in the literature on tourism experience, particularly in the context of the advent of mobile technologies and the option for tourists to share instantly their experiences to their social networks (Jansson, 2007; Kirillova and Wang, 2016; Fan *et al.*, 2019). Although this literature approaches the experience of the tourist and not specifically the experience of the SMI, the analysis of how the travel experience is retold is, nevertheless, relevant to our study, since the retelling of the experience is a key aspect of the influencer's job. In the following, we will look further into how the retelling of the touristic experiences of the SMIs can be explored through the concepts of en- and decapsulation.

André Jansson (2007) frames the experience of being a tourist as a division between being at home and away – between the ordinary and the extraordinary (Jansson, 2007). In the 'classical style' of tourism, the distinction between the travel experience and everyday life, or between 'work time' and 'leisure time', is reinforced by photographs, films, postcards and souvenirs that are recorded or acquired *during* the experience, and then brought home and 'unpackaged' or retold *at home* to an audience of family or friends (Jansson, 2007). According to Jansson, this accentuates the social, cultural and geographical distance between both the physical and psychological space of 'home' and 'away' (Jansson, 2007). The author describes the process of recording or acquiring mementos as 'encapsulation', and the subsequent sharing of mementos as 'decapsulation', which in itself underlines the notion that the gathering of stories and experiences should happen independently from their retelling, with 'encapsulation' being completed in a time and space that is totally separate from the time and space in which 'decapsulation' occurs.

However, Jansson argues that with the introduction of technologies that allow immediate, online sharing the encapsulation and decapsulation process is problematized and its 'magic' is eliminated (Jansson, 2007). According to the author, mobile, networked media blur the boundaries between physical and imagined mobility, and between everyday life and the tourism experience (Jansson, 2007). Added to this, Jansson argues, there are nowadays greater obligations for tourists to be reachable while on holiday (Jansson, 2007), which perhaps encourages even more immediate sharing.

These are certainly valid points, although the context is already outdated. At the time of writing, in 2007, the mobile, networked media of which Jansson wrote were online blogs and text messages. He also mentions the advent of digital photography and video which newly enabled tourists to delete, re-record, and edit their visual material, and thereby manipulate the retelling of their experiences to a greater extent (Jansson, 2007). Images, movies and, of course, text could be shared immediately via SMS or MMS, and recipients could give direct feedback and interact in the retelling of the trip while it was still ongoing (Jansson, 2007). These developments, which to Jansson seemed to be a revolution of new technologies in themselves, are nowadays seen as the 'tip of the iceberg' of a digital revolution. In this process, the boundaries of home and away were blurred to a much greater extent. However, this does not invalidate Jansson's arguments; on the contrary, they can be seen to be even more relevant in today's context.

Today, smartphones and faster, more wide-reaching internet connections allow the immediate uploading and sending of visual materials, and social media allow not just friends and family but larger social networks to give immediate feedback on this communication (Gretzel, 2017). New technologies have become increasingly more portable, wearable and affordable, and opportunities that were previously available only to professional photographers and filmmakers, such as the option to record high-quality aerial video, are now open to the everyday traveller (for example, with the availability of cheap and portable drones) (Gretzel, 2017). In the context of the 'retelling' of travel experiences, these developments, arguably, allow travellers and especially SMIs to be much more expressive in their 'decapsulation', and for their content to be more immersive, perhaps further diminishing the distance between the everyday life of the audience and the travel experience.

Another development that can link specifically to SMIs are the front-facing cameras on smartphones and selfie sticks (Gretzel, 2017), which have, arguably, contributed to a shift in focus in the 'encapsulation' of experiences, as travellers and content creators are encouraged to highlight images and videos of themselves, or themselves in the destination. In the case of SMIs, this makes it easier for the SMIs to build a persona and a connection with their audience, since they can talk directly to the camera. Since 2015, SMIs (and other users) can even talk to their followers in real time, with live-streaming applications that have become integrated into many social media platforms, such as Facebook and Instagram (Gretzel, 2017).

In the decade or so since Jansson's article began describing a significant turn in the encapsulation and decapsulation of tourist experiences, a number of scholars have addressed advances in media technologies and their effects on tourists' gathering and retelling of experiences (Gretzel, 2017; Seeler *et al.*, 2019; Fan *et al.*, 2019). It is generally agreed that this storytelling nowadays occurs digitally, in real time, and in a two-way fashion (Gretzel, 2017; Seeler *et al.*, 2019; Fan *et al.*, 2019). In other words, the decapsulation of the experience is co-created between the narrator and the listener (Seeler *et al.*, 2019). Another aspect worth mentioning in this regard is the increased control that tourists now have in retelling their travel stories. Users now have the freedom and autonomy to decide when, where and how to decapsulate their experience (Fan *et al.*, 2019). Physical and geographical factors seem to have increasingly less significance.

10.2.5 Contribution of this study

One of the primary ways in which our case takes this literature further is that tourism SMIs themselves represent an intriguing blurring of boundaries between home and away. Since professional tourism SMIs are often paid or at least sponsored to travel to destinations and retell their experiences, the blurring of the spheres of home and away is potentially even more apparent, since they are simultaneously at work and acting as tourists. Furthermore, the social media platforms which have previously been argued to blur these boundaries are exactly where SMIs carry out their work. It is easy to imagine that there is an interesting tension at play between being connected, being on holiday, being at work, being at home, and being in Greenland.

The closest that the literature has come to examining this relationship so far is Fan *et al.* (2019), who have contributed to bringing Jansson's (2007) work up to date, and to constructing a typology of tourists according to the extent to which they engage with social media while on holiday. The typology classifies tourists, rather than SMIs, but the authors' 'dual-zone traveller' is perhaps the closest academic analysis of the role of an SMI in tourism along with the tension between being connected and being away. This is how the authors describe a dual-zone traveller: 'This group is determined to manage their original and away worlds simultaneously. They keep a high level of interaction with their original social network, without abandoning the opportunity to have contact with local people and their travel companions' (Fan *et al.*, 2019, p. 1010). The authors describe a traveller who manages to skillfully and simultaneously balance their 'home' and 'away' worlds, and who can easily move 'out of' one world and into the other (Fan *et al.*, 2019, p. 1010). They argue that these types of tourists can only immerse themselves in each world to a limited extent because they need to be able to move to the other zone at a moment's notice (Fan *et al.*, 2019, p. 1010). Although more specific literature analysing the role of an SMI in this regard is lacking, this is a useful starting point for our discussions.

Although the concepts of encapsulation and decapsulation were updated according to recent technological developments by Jansson (2007) and others since then (Fan *et al.*, 2019), it is clear that they require updating again, as technological developments and users' roles and relationships to these technologies evolve. The following analysis and discussion will take the existing literature further by examining how these concepts are negotiated in a context in which connectivity can no longer be taken for granted. It will also contribute by examining the 'SMI experience' in tourism for the first time, with a focus specifically on how the respondents negotiate the blurred boundary between home and away, online and offline, experience-gathering and experience-retelling.

10.3 Methodology and Case Description

In this chapter, we are interested in unravelling the micro-practices of social media marketing through the perspective of professional social media influencers. We explore what such a practice might look like in an Arctic environment such as Greenland, framed within a context of geographical peripherality, and a place where digital connectivity is limited and expensive. We use the case of Greenland to trace the attempts of the influencers to balance and make sense of Arctic digital connectivity. By doing so, we wish to discuss the Arctic both as a place of extremes and as off-grid, and as a mundane and globally connected destination. Also, by exploring the fine-grained processes and negotiations related to social media marketing in this remote area, we hope to open up related issues, which are less observable in well-connected destinations, where staying connected and being online is taken for granted.

Greenland, unlike many other tourist destinations (especially in the western world) is not covered by a reliable internet connection. Internet is expensive, limited and often slow, meaning that even those businesses that have WiFi will often charge guests to use it. Furthermore, even if tourists can afford to pay for WiFi and mobile data, there are many remote locations in Greenland that simply do not have a mobile data connection. This, then, represents a potential complication in that users, and especially those users who rely on an internet connection (such as SMIs), perhaps have less control over the decapsulation of their experience than they are used to. The context of Greenland provides an interesting inversion or perhaps reversal of the norm. As a result, the question is raised as to what impact this context has on an SMI (and the resulting retelling of the SMI's experience) that is not just used to but, to some extent, depends on a reliable and continuous internet connection.

As the national destination marketing and management organization, Visit Greenland is the main body responsible for marketing Greenland to a global public. An element of Visit Greenland's social media marketing mix involves sponsoring appropriate and prominent social media influencers to travel to Greenland and post about their experiences. For this study, we use one of these so-called 'influencer trips' as a case.

Since much existing literature analyses influencer posts by searching only for that content which is hashtagged '#sponsored' or tagged with 'paid partnership', it naturally ignores more subtle influencer content which is not tagged. In the case of this study, the role of the second author as an industrial PhD employed at Visit Greenland offers a unique insider perspective, providing knowledge of Greenland's SMI content which is sponsored but untagged.

The influencer trip that constitutes the empirical data for this study took place in early March 2020 and was intended to promote Greenland's primary destinations in the winter season. It lasted five days and four nights, with two nights spent in Nuuk, the capital city, and two nights in Ilulissat,

Table 10.1. The trip participants.

Name	Instagram handle	No. of followers (Instagram)	Origin market	Top 3 audience markets
Luke	withluke	363,000	UK	Western Europe, UK, North America
Kyle	kylevollaers	108,000	South Africa	Western Europe, UK, North America
Peter	yantastic	284,000	Australia	North America, Australia, western Europe
Natasha	thebohochica	32,500	Middle East	Middle East
Asa	asasteinars	265,000	Iceland	Iceland, western Europe, North America

the country's third largest town, and the surrounding area. Both are popular summer destinations; the aim behind pushing them as winter destinations is to help extend the tourism season into the 'shoulder seasons', a commitment voiced in the 2020 strategy of Visit Greenland. The trip was planned to give the participants an overview of the highlights of the two destinations: they experienced a city tour, a boat tour, local food and a snowshoe hike in Nuuk, and in Ilulissat, dogsledding and an Arctic sauna experience. A planned iceberg sightseeing boat tour in Ilulissat was cancelled due to an unseasonal amount of ice in the fjord making it impossible for boats to sail. The first night in the Ilulissat area was spent in the wilderness at 'Igloo Lodge', where the tourism operator World of Greenland has built snow igloos for guests to spend the night in. The lodge is reached by snowmobiles and there are tours included in the 24-hour-long experience.

The five influencers were chosen from Visit Greenland's 'media request' applications, which is a rolling application system allowing anyone to pitch an idea for a sponsored trip. Two of the influencers (Luke and Kyle) applied together, while the other three all applied individually. The influencers were grouped together to save time and money on the co-ordination of the trip, and because Visit Greenland at the time did not have the resources to send a guide with them for the duration of the trip. In order to avoid similar content being sent to the same markets and audiences, Visit Greenland tried to select influencers who had different styles and audiences in different target markets. See Table 10.1 for information about each of the trip participants.

For the purpose of collecting empirical material for this chapter, telephone interviews were conducted with the five influencers shortly after their return from Greenland. Three of the interviews were conducted individually and one was conducted with two of the SMIs together (Luke and Kyle). The interviews lasted, on average, 30 minutes each and were all audio-recorded. Conversations focused primarily on the SMIs' specific experiences as professional social media influencers in Greenland, on what was both positive and challenging about combining a remote travelling experience with the expectation to regularly update their online audiences. The interviews were subsequently transcribed and thematically analysed, manually, by consulting with the literature reviewed above. This led to the identification of three themes: being in the moment; digital connectivity and 'liveness'; and home/away distinctions, all of which we elaborate on in the analysis below.

10.4 Analysis

10.4.1 Greenland – an altering experience

Greenland is often – in tourism research and in destination marketing – narrated as out of the ordinary and as a huge and memorable experience. While this might not be an unusual marketing approach, Weaver and Lawton (2017) argue that 'Even by the exotic standards of the Arctic as a region, Greenland (or Kalaallit Nunaat) stands out for its impressive idiosyncrasies' (p. 56). In our conversation with the influencers, this 'extraordinarity' was indeed apparent. While all of the influencers are avid and well-accustomed travellers, all expressed, even spontaneously, the awe or surreal feeling of being in Greenland. Peter expressed this in the following way: 'I think for me even like now, I look back, I just can't believe that I was in Greenland.'

Some saw the experiences in Greenland as amplified because of a lack of expectations, which Luke explained in the following way: 'Perhaps we don't see it much on social media, like the whole region, so we're not sure what to expect.' While Greenland's exceptionalism is usually explained through geographical peripherality and limited access, this quote suggests that Greenland is also situated in what we might term 'digital peripherality', in which Greenland and the Arctic regions lack presence and representation within the social media destination imagery.

As a professional and active co-producer of destination imagery, Asa gives us an indication of how digital peripherality comes about in explaining the paradox of wanting to share and mediate the Greenlandic experience in the following way: 'I mean it's so cool to come to Greenland, it's so remote, it's so extreme. And then you just don't have data.' These quotes suggest a situation where social media travel influencers visit a destination with amazing imagery but are in some cases challenged in sharing content. As we shall now see, this required some negotiation between online absence and presence, but also offered new ways of engaging with followers and telling stories.

10.4.2 Being in the moment – (re)discovering off-line encapsulation

In perhaps the most 'altering' part of their trip, the influencers were taken on an overnight stay to an igloo lodge situated 30 minutes' drive on snowmobiles from Ilulissat, the primary tourist destination in Greenland. Ilulissat is best-known for its glacier, claimed to be the world's fastest-moving, which sends colossal icebergs into the bay surrounding the town. Aside from the icebergs, Ilulissat is Greenland's third largest town and it offers a comprehensive cultural experience. As it sits above the Arctic Circle, it is also attractive for dogsledding and other winter activities.

The Igloo Lodge experience is 24 hours long and includes a snowmobile journey to and from the lodge, local food experiences in the hut, and either a snowshoe hike or a snowmobile tour in the wilderness surrounding the lodge. Although the intention is that guests will spend the night in the snow igloos, there is also a heated hut for them to sleep in if it gets too cold.

All of the influencers mention their stay in the Igloo Lodge as something out of the ordinary. Much of what they recounted was connected to being offline for a longer period of time (24 hours, overnight): 'You do find yourself being in the present moment a lot more because you're not constantly connected' (Natasha). To influencers, whose work (and often also social) life is digitally enabled, a break in connectivity is an unusual thing. Natasha elaborates on what that entailed:

> 'For me it was a positive thing not to have internet there [at the lodge]. Just being in the lodge and talking to the people who stay at the lodge, being outside, seeing the northern lights. I think it was nice not to have connectivity there...one night just being there and being in the moment and embracing the whole thing about being somewhere so far away.'

As Peter exemplifies in the next quote, influencers even found themselves 'discovering themselves' being disconnected:

> 'I think one of my stories I did specifically say, "Hey it was such a refreshing or nice thing to be disconnected from the world for 24 hours. You know, I've never done that before." I think when you're there you automatically put your hand in your pocket, take your phone out, but then you realize that there's no internet, and you put it back [laughs] and you go back to the conversation.'

The Igloo Lodge experience shows how highly mobile and digital travellers (re)discover a new way of being and dwelling in travelling, as further developed by Peter:

> '...being completely cut out from the rest of the world for 24 hours. So that was a new thing for all of us...at night, we basically sat around and, you know, really talked, had conversation around the candles, which was really cool. You know usually you have people just sitting around on their phone all the time. But I think the positive side of that is people having that communication, you know, having real conversations, face-to-face, having that eye contact, you know, back to the really old days.'

This disconnectivity sparked reflections about 'being in' and 'embracing' the moment, nature and the conversation. It created a situation similar to that in 'the really old days' where the process of recording or storing mementos was decoupled from the decapsulation, which for these influencers is usually possible everywhere, constantly. So 'you put your phone back in your pocket,' as argued by Peter. While the above quotes display an almost idyllic situation for the influencers on a personal level, as they seem to experience a refreshing encounter with presence and the 'old days', the interviews unravelled other, professional, reflections triggered by the lack of connectivity, and, as shown in the following, these were not entirely without friction.

10.4.3 Negotiating online connectivity: on or off?

Despite enjoying a night of candle-lit conversations with eye contact and 'old-school presence', Kyle comments on some of the downsides of the night in the Igloo Lodge:

> 'For social media influencers or for what we do, we wanna be able to check in from time to time, right? So when we come back from our activity we'll wanna quickly go on social media and update and just check in with our audience, post videos and all that. And that night we saw northern lights, you know, it would have been great to do a live story – "Hey guys, look at this", you know, bla bla bla. So I think that's a specific moment I wanted to have internet access. But obviously, being in the middle of nowhere, we couldn't. And also we couldn't charge our phones or charge our batteries either because there was no power there, so that was a challenging part as well.'

The professional downsides of being offline were also experienced by Luke, as he describes here:

> 'We really wanted to do a live story on Instagram. And I said to Kyle, let's do a live story. Do you know what a live story is? It's like Facebook Live, where me and Kyle could join as a group, and all our followers are notified that we're going live. Because we wanted to share from that location a really cool moment. But the problem was we had no 4G or anything so we couldn't. So that was a negative.'

The importance of 'going live', as a tourism SMI, is expanded on by Peter: '[Instagram Live] is a great way to engage with your audience, for them to see exactly what you're doing and what you're seeing. For you to capture the moment, you know, live. I think that's a good thing as well.' Liveness, and thereby the feeling of connection, is an integral part of the influencer–follower relationship.

While the above statements point to the felt necessity to be online and connected to an audience – to have constant (potential) online presence – not all SMIs perceived this as an ideal. While agreeing on the importance of liveness, Natasha notes that it is necessary to be selective about when you 'go live' as an SMI: 'I need to be connected every day because that's how my audience...they're a part of the journey, they're experiencing all of these things with me. But at the same time it's not necessary for me to be live all the time because sometimes that can take away from what I'm doing in the present moment.' To Natasha, a balance exists between being 'present' in the moment, with herself, and being 'live' on social media, with her followers. She suggests here that she cannot be 'present' and 'live' at the same time, and that she must, therefore, negotiate her movement between the two states.

What also emerges from Natasha's explanation is a difference between the concepts and purposes of being connected and being live. Being, or going, live is something that she consciously manipulates depending on the situation and her need (or lack thereof) to be present. Being connected, however, is something that she requires more constantly – 'every day' – in order to make her audience feel that they are a part of the journey. In other words, connectivity is a necessity, whereas liveness is more of an added extra, when relevant. Of course, being connected is also a prerequisite to being able to go live.

Connectivity as a necessity is also touched on by Luke, when he recalls: 'I posted a photo on the morning before we went out [to the Igloo Lodge] and then I couldn't engage with it. And it didn't do so well because of that. Because you're away from WiFi, and we couldn't reply to messages and things and also even e-mails in terms of brand work and stuff like that.'

Kyle goes on:

> 'Luke and I, we have a sort of a specific time that we post every day. And quite often we couldn't make the time because we had to leave to go somewhere, and the time difference, the time zone, made it tough. Because we didn't have a connection on the road so we weren't able to post on the go. We missed our peak posting time, which then, when we don't post at those good times, our engagement is far lower for the posts.'

For Luke and Kyle, being connected is also more of a fundamental part of their professional experience than going live. The ability to respond to and engage with reactions to a specific piece of content is something which affects the success of that piece of content. Kyle points out a temporal element to this connectivity, suggesting that it is not simply a professional requirement for them to be connected 'every day' but at specific times of every day, corresponding to their 'peak posting time'. Posting at this peak time is also crucial to the success of their content. In the following and final section of our analysis, this temporal element will be explored further and in relation to the tension introduced by Natasha, between the states of 'being present', being connected and 'going live'.

10.4.4 Home and away – a blurred or necessary distinction?

In this part of the analysis, findings are brought together to discuss how they impact and complicate each other, and how they further the issues presented in our literature review. Natasha's allusion to different 'states' (being present, being connected, and going live), and her ability to move consciously and appropriately between them, is reminiscent of Fan *et al.*'s (2019) description of the 'Dual-zone traveller': a traveller who skillfully balances their 'home' world and their 'away' world, and can move between them almost instantly. However, the 'worlds' she balances are, arguably, not accurately described as 'home' and 'away', but rather as different spaces, or different modes of being 'there'. She talks of being 'there' as being present, in the moment, and being 'there' as being connected to and available for her followers.

We could argue, therefore, that the worlds she moves so skillfully between are the physical place (presence in the moment) and the digital space (being connected online). Bearing in mind that Fan *et al.*'s (2019) typography is a typography of tourists, and so describes the tourist experience, we argue that, when analysing the tourism SMI experience, the distinctions of 'home' and 'away' are not appropriate. The decapsulation process for a tourism SMI does not represent communication with a 'home' world, because their audience is not family and friends – it is their followers, with whom they have a professional relationship. The blending of professional and personal elements in this way makes the tourism SMI experience a novel and intriguing topic of study.

In turn, this complicates Jansson's (2007) arguments, which Fan *et al.* also used as a theoretical foundation to develop their typography. Jansson argued that new technologies blur the boundaries between encapsulation and decapsulation, removing its 'magic', since travellers could continually 'decapsulate' during their journey (Jansson, 2007). However, what we see in the case of our SMIs is more complex than this. The fact that Natasha sees herself as moving between physical and digital states suggests that there is a conscious encapsulation and decapsulation process occurring, albeit much more frequently than in the 'traditional' manifestation of the process. However, Natasha's clear differentiation between the states of being present (where she encapsulates) and being connected (where she decapsulates) indicates that there is not necessarily a blurring of boundaries here; on the contrary, the states are distinctive and have markedly different purposes. We argue, therefore, that the experience of the SMIs does not represent a blurring of boundaries between encapsulation and decapsulation, as Jansson described, but rather a change in the way in which the travellers decapsulate. The different states are still well defined and mutually exclusive, according to the experiences of our respondents.

Bringing in the temporal element of the SMIs' connectivity, as touched on by Kyle, we can argue that, although the encapsulation and decapsulation process happens frequently, it happens (or ideally so) according to a schedule which is predefined by the SMIs. This is a schedule determined to result in the best engagement for their posts. Through adhering to their 'peak posting time', they suggest that they have built up a relationship with their followers based on the expectation that the SMIs will post at particular times, and engage with their audience immediately following the act of posting. This reinforces the argument that the SMIs consciously move between the physical place and the digital space in what we might see as a 'rhythm of connectivity', unique to each SMI and dependent on different factors.

For our respondents, decapsulation has to take place along their individual rhythm of connectivity for optimal performance. This presents challenges and potential frustration in a destination like Greenland, since our SMIs could not always be sure that they would have an internet connection at the time at which they were scheduled to decapsulate. On the other hand, this frustration was reinterpreted in some cases by the SMIs as an opportunity to create drama. For example, at the end of his trip in Greenland, Kyle posted a story where he writes: 'I just spent the past few days off-grid', which is an exaggeration, since he was only technically 'off-grid' for 24 hours, during the Igloo Lodge trip.

While he presumably realizes the exaggeration of this statement on the remoteness and lack of connectivity of his trip, he also adds to its intrigue and 'otherness' for his audience. In this way, Kyle leverages his lack of connectivity to enhance his storytelling. He then goes on to decapsulate his trip in the stories which follow this one, in a technique reminiscent of the 'traditional' encapsulation and decapsulation process about which Jansson (2007) was so nostalgic. This suggests that the classical method of encapsulation and decapsulation is something to which (even young and modern) audiences still relate and recognize, and which can nowadays be leveraged in a dramatic way, since it creates a tension and suspense to which audiences (particularly those followers who are attuned to particular rhythms of connectivity) are not accustomed.

10.5 Discussion and Conclusion

In this chapter we examined the SMI experience in tourism for the first time, with a focus specifically on how the respondents negotiate the blurred boundary between home and away, online and offline, experience-gathering and experience-retelling. Our conversations with the tourism SMIs displayed the trade-offs and balancing acts of being 'online' and 'off the beaten track', seeing them as equally important but at times hard to reconcile. Also, they generated an understanding of the constant intersecting and enmeshing of professional and personal practices of the SMIs, thus reflecting and magnifying three global tendencies in tourism.

Firstly, the general understanding within the mobilities turn of tourism acting not as 'an ephemeral aspect of social life that is practised outside normal, everyday life' but rather 'as integral to wider processes of economic and political development processes and even constitutive of everyday life (Hannam *et al.*, 2014, p. 172). During the trip, the SMIs were simultaneously working, exploring and reflecting on a personal as well as professional level. The many sides of the experience, as work and as leisure, as a personal and as a professional undertaking, were not clearly compartmentalized, but rather blurred or perhaps rather seamlessly performed into a coherent whole, thus bringing empirical support for the theoretical claims within the mobilities turn.

Secondly, this blurry outcome of what we might call the tourism SMIs' practice points to another dominant and growing trend, namely the increasingly technologically mediated character of tourism products in the experience economy (Tussyadiah and Fesenmaier, 2009). It is this development which has enabled the tourism SMI profession to begin with, while the production of SMI tourism marketing reversely strengthens the consumers' ability to identify and familiarize themselves with destinations such as Greenland, which, as argued, can be difficult to explore and access online.

Thirdly, the practices of our SMIs underscore the increasing co-production of tourism experiences and products (Prahalad and Ramaswamy, 2004) and the co-created nature of social media content. The social media content posted and shared by the SMIs is not prefabricated products, but is decapsulated in a collaborative process together with followers. Thus, social media influencers and consumers as potential tourists co-produce content through interactions, engagement, sharing, and liveness, all of which depend on connectivity.

10.5.1 Implications

While it is (obviously) always the case that online sharing is dependent on digital connectivity, using Greenland to disrupt and slow down the usually 'smooth' online practices of SMIs enabled us to explore these practices in more depth. Recalling their trip to Greenland, SMIs highlighted and reflected upon the constant negotiation of especially three parts of their social media practices: the *blurring* of work/leisure and personal/professional, the *technological mediation* and the *co-creation* of the product. To qualify this negotiation, we identified different rhythms of connectivity among the SMIs, suggesting that good and constant connectivity is not always a necessary condition for travelling and sharing 'off the beaten track' adventure destinations such as Greenland.

As the current chapter has explored the touristic experience of SMIs, which has been identified as a research gap, the development of social media platforms and usage continues to develop in unprecedented and unexpected ways. Concurrently, digital infrastructures continue to improve, constantly improving global online connectivity, also in remote areas and destinations. While challenges and frustrations about internet access might foreseeably become reduced over the coming years as a consequence of this, the identification of certain individual rhythms of connectivity might be relevant to think about from an industry perspective in that, as a DMO, it becomes valuable to consider the individual rhythm of connectivity of an SMI with whom one is working.

This implication applies not just to remote and less-connected destinations such as Greenland, but also to any destination which employs partnerships between DMOs and SMIs as a marketing strategy. Learning about an SMI's unique rhythm of connectivity and planning the trip in accordance with this would allow, for example, the SMI to have downtime at their peak posting time, so that their posts could achieve maximum engagement and the destination receives optimal exposure. In this way, it is interesting to think about how co-creation between the DMO and SMI can already play a part in the development of the trip itinerary, before the trip has even begun.

It is also worth noting that, as well as rhythms of connectivity being unique to each SMI, as argued above, they will also change according to the destination. This contextual element has been highlighted in this study by Kyle's anecdote about being unable to share his northern lights experience live as it was happening. For DMOs which want to highlight time-dependent experiences such as this, therefore, it is advisable to incorporate these considerations into the planning of the trip, to ensure connectivity at the relevant times.

10.5.2 Limitations and future research

This research represents one of the first ventures into the academic analysis of the SMI experience. Although it is somewhat pioneering in this manner, it also provides only a limited insight into the field. Since this study has revealed the numerous tensions and complexities of the SMI experience, future research is encouraged which delves into these intricacies in more depth and in other destinations. Given the speed at which the digital world and the social media landscape is transforming, there is great potential for future research in this area.

References

Bjørst, L.R. and Ren, C. (2015) Steaming up or staying cool? Tourism development and Greenlandic futures in the light of climate change. *Arctic Anthropology* 52(1), 91–101. DOI: 10.3368/aa.52.1.91.

Buhalis, D. and Law, R. (2008) Progress in information technology and tourism management: 20 years on and 10 years after the Internet – the state of eTourism research. *Tourism Management* 29(4), 609–623. DOI: 10.1016/j.tourman.2008.01.005.

Chae, I.Y., Lee, Y.M., Yu, K.Y. and Kim, J.Y. (2017) A method for analysis of preferences of places using social media text. *Journal of the Korean Society for Geospatial Information Science* 25(4), 55–64. DOI: 10.7319/kogsis.2017.25.4.055.

Cuevas, L.M., Chong, S.M. and Lim, H. (2020) Influencer marketing: social media influencers as human brands attaching to followers and yielding positive marketing results by fulfilling needs. *Journal of Retailing and Consumer Services* 55, 102–133.

De Jans, S., Van de Sompel, D., De Veirman, M. and Hudders, L. (2020) #Sponsored! How the recognition of sponsoring on Instagram posts affects adolescents' brand evaluations through source evaluations. *Computers in Human Behavior* 109(1), 106342. DOI: 10.1016/j.chb.2020.106342.

De Veirman, M., Cauberghe, V. and Hudders, L. (2017) Marketing through Instagram influencers: the impact of number of followers and product divergence on brand attitude. *International Journal of Advertising* 36(5), 798–828. DOI: 10.1080/02650487.2017.1348035.

Dhanesh, G.S. and Duthler, G. (2019) Relationship management through social media influencers: effects of followers' awareness of paid endorsement. *Public Relations Review* 45(3), 101765. DOI: 10.1016/j.pubrev.2019.03.002.

Fan, D.X., Buhalis, D. and Lin, B. (2019) A tourist typology of online and face-to-face social contact: destination immersion and tourism encapsulation/decapsulation. *Annals of Tourism Research* 78(3), 102757. DOI: 10.1016/j.annals.2019.102757.

Farivar, S., Wang, F. and Yuan, Y. (2019) Meformer vs. informer: influencer type and follower behavioral intentions. *Proceedings of Twenty-fifth Americas Conference on Information Systems*, Cancun.

Fäs, J. and Zumstein, D. (2019) Loyalty programs in tourism: how influencers and comminities change word-of-mouth marketing. *Proceedings of the 18th International Conference on WWW/Internet*.

Freberg, K., Graham, K., McGaughey, K. and Freberg, L.A. (2011) Who are the social media influencers? A study of public perceptions of personality. *Public Relations Review* 37(1), 90–92. DOI: 10.1016/j.pubrev.2010.11.001.

Gretzel, U. (2017) The visual turn in social media marketing. *Tourismos* 12.

Hannam, K., Butler, G. and Paris, C.M. (2014) Developments and key issues in tourism mobilities. *Annals of Tourism Research* 44(3), 171–185. DOI: 10.1016/j.annals.2013.09.010.

Hays, S., Page, S.J. and Buhalis, D. (2013) Social media as a destination marketing tool: its use by national tourism organisations. *Current Issues in Tourism* 16(3), 211–239. DOI: 10.1080/13683500.2012.662215.

Jansson, A. (2007) A sense of tourism: new media and the dialectic of encapsulation/decapsulation. *Tourist Studies* 7(1), 5–24. DOI: 10.1177/1468797607079799.

Kaplan, A.M. and Haenlein, M. (2010) Users of the world, unite! The challenges and opportunities of social media. *Business Horizons* 53(1), 59–68. DOI: 10.1016/j.bushor.2009.09.003.

Kaplan, A.M. and Haenlein, M. (2012) The Britney Spears universe: social media and viral marketing at its best. *Business Horizons* 55(1), 27–31. DOI: 10.1016/j.bushor.2011.08.009.

Kirillova, K. and Wang, D. (2016) Smartphone (dis)connectedness and vacation recovery. *Annals of Tourism Research* 61(1), 157–169. DOI: 10.1016/j.annals.2016.10.005.

Lee, Y.S., Weaver, D.B. and Prebensen, N.K. (2017) Arctic destinations and attractions as evolving peripheral settings for the production and consumption of peak tourism experiences. In: *Arctic Tourism Experiences: Production, Consumption and Sustainability*. CAB International, Wallingford, UK, pp. 1–8.

Lin, R. (2018) Silver lining of envy on social media? The relationships between post content, envy type, and purchase intentions. *Internet Research* 28(4), 1142–1164. DOI: 10.1108/IntR-05-2017-0203.

Magno, F. and Cassia, F. (2018) The impact of social media influencers in tourism. *Anatolia* 29(2), 288–290.

Minazzi, R. (2015) *Social Media Marketing in Tourism and Hospitality*. Springer.

Omnicore Agency (2019) Instagram by the numbers (2019): stats, demographics & fun facts. Available at: https://www.omnicoreagency.com/instagram-statistics/ (accessed 22 November 2019).

Parise, S., Guinan, P.J. and Weinberg, B.D. (2008) The secrets of marketing in a web 2.0 world. *The Wall Street Journal* 4.

Phua, J., Jin, S.V. and Kim, J.J. (2017) Gratifications of using Facebook, Twitter, Instagram, or Snapchat to follow brands: the moderating effect of social comparison, trust, tie strength, and network homophily on brand identification, brand engagement, brand commitment, and membership intention. *Telematics and Informatics* 34(1), 412–424. DOI: 10.1016/j.tele.2016.06.004.

Prahalad, C.K. and Ramaswamy, V. (2004) Co-creation experiences: the next practice in value creation. *Journal of Interactive Marketing* 18(3), 5–14. DOI: 10.1002/dir.20015.

Qualman, E. (2012) Social Media Video 2013. Available at: https://www.youtube.com/watch?v=QUCfFcchw1w (accessed 22 November 2019).

Schegg, R., Liebrich, A., Scaglione, M. and Ahmad, S.F.S. (2008) An exploratory field study of Web 2.0 in tourism. In: O'Connor, P., Hoepken, W. and Gretzel, U. (eds) *Information and Communication Technologies in Tourism 2008*. Springer, Vienna, pp. 152–163.

Seeler, S., Lück, M. and Schänzel, H.A. (2019) Exploring the drivers behind experience accumulation: the role of secondary experiences consumed through the eyes of social media influencers. *Journal of Hospitality and Tourism Management* 41(4), 80–89. DOI: 10.1016/j.jhtm.2019.09.009.

Statista Market Forecast (2019) Social media advertising. Available at: https://www.statista.com/outlook/220/100/social-media-advertising/worldwide (accessed 5 June 2019).

Statistics Greenland (2019) Greenland in figures. Available at: http://www.stat.gl/publ/en/GF/2019/pdf/Greenland%20in%20Figures%202019.pdf (accessed 28 May 2020).

Stoldt, R., Wellman, M., Ekdale, B. and Tully, M. (2019) Professionalizing and profiting: the rise of intermediaries in the social media influencer industry. *Social Media + Society* 5(1), 205630511983258. DOI: 10.1177/2056305119832587.

Tussyadiah, I.P. and Fesenmaier, D.R. (2009) Mediating tourist experiences: access to places via shared videos. *Annals of Tourism Research* 36, 24–40.

Weaver, D.B. and Lawton, L.J. (2017) Degrees of peripherality in the production and consumption of leisure tourism in Greenland. *Arctic Tourism Experiences: Production, Consumption and Sustainability*, CAB International, Wallingford, UK, pp. 56–66.

Xiang, Z. and Gretzel, U. (2010) Role of social media in online travel information search. *Tourism Management* 31(2), 179–188. DOI: 10.1016/j.tourman.2009.02.016.

Xu, X. and Pratt, S. (2018) Social media influencers as endorsers to promote travel destinations: an application of self-congruence theory to the Chinese Generation Y. *Journal of Travel & Tourism Marketing* 35, 958–972.

11 Promoting Intangible Cultural Heritage through Social Networks: A Case Study of the *Fête De l'Ours* in France

Jordi Arcos-Pumarola[1]* and Marta Conill-Tetuà[2]

[1]*CETT-UB Barcelona School of Tourism, Hospitality and Gastronomy, Barcelona, Spain*
[2]*Consell Comarcal del Vallès Occidental, Terrassa, Spain*

11.1 Introduction

Expressions of Intangible Cultural Heritage (ICH), such as traditions, dances and social performances, play a vital role in shaping the identity and image of a territory. These cultural expressions are inherited from the past; that is, they have passed from generation to generation, carried forward by individuals who protect and transmit these practices (UNESCO, 2018). This process of inheritation is one of the main features of ICH; its preservation and protection depend on the present community embracing and reproducing the traditional cultural practices related to ICH. This transmission process is justified because, as Ruiz (2020) shows for the case study of Valencia's Fallas, ICH is valuable for the communities who see it as an element in the understanding and staging of their cultural background, as well as an asset to differentiate themselves from other communities.

This capacity of ICH to single out territories is particularly interesting for the tourism industry, which understands ICH as a useful cultural resource for the transmission of cultural authenticity and, in this way, may foster the competitivity of a destination (Kim *et al.*, 2019). However, the use of ICH in the context of the tourism industry must be addressed, since its commercialization is sometimes linked to negative consequences, such as unequal distribution of tourism benefits, cultural appropriation of ICH by external agents or appearance of conflicts within the communities, among others (Salvador and Abellan, 2020).

*Corresponding author: jordi.arcos@cett.cat

DOI: 10.1079/9781789248753.0011

In parallel, in recent years, the importance of social networks in tourism marketing has grown. Social networks allow us to share our experiences at any stage of travel, as well as explain those experiences to anybody who visits our profile. In this way, user-generated opinions shared on social platforms affect future travellers' perceptions of destinations (Narangajavana Kaosiri *et al.*, 2019). As a result, social networks have surpassed their origins as merely spaces to share images, experiences and thoughts, to become promotional platforms and a useful tool for marketing and selling. The transformation of social networks into promotional platforms is particularly relevant in the case of the tourism industry, given the importance of information in the process of selecting a destination or booking a product (Hays *et al.*, 2013).

Considering the cultural relevance of ICH and its meanings for communities, and the fact that it is increasingly becoming an asset of destinations in promoting tourism, the use of ICH in social networks must be put in the spotlight to identify best practices of transmitting the meanings of ICH through social networks. An analysis of the use of ICH in the context of social networks will also illuminate which particular tourist image is being transmitted by the various agents and stakeholders that interact with social networks. In this vein, this chapter will present a case study based on the *Fête de l'Ours*, a celebration performed in three rural villages in the south of France, looking at how it is explained within a particular social network, namely, Instagram. In this way, the case study will aim to: (i) identify the elements of the festivity that are highlighted by locals and visitors in their *Fête de l'Ours* account on Instagram; and (ii) analyse how locals, visitors and stakeholders explain and share the experience of the festival.

11.2 Literature Review

11.2.1 The role of intangible cultural heritage in the tourism context

The evolution of the concept of heritage during the last century has caused ICH's growing recognition as a valuable asset of communities (Vecco, 2010). According to Su *et al.*'s (2019) bibliometric analysis of ICH-related literature, communities understand ICH as valuable for four main reasons. First, ICH has a historical value, allowing first-hand access to the history of a particular community. Second, it illuminates how a particular society understands itself, what its values and forms of knowledge are; in this sense, the preservation of ICH expressions is particularly relevant in fostering intercultural dialogue in an increasingly globalized context (UNESCO, 2018). Third, ICH provides an emotional and aesthetic cultural experience, enriching people's lives, and reinforcing their links and ties with the place. Finally, ICH offers potential economic value and communities can develop strategies to generate benefits around it.

One of the most common uses of ICH as an economic resource is found in the context of cultural tourism. Given the close and exclusive relationship of ICH with its territory, the tourist industry is well aware of the potential of ICH and its capacity for singularization. This way, ICH has been a key actor in the progressively broadening process of cultural tourism (McKercher, 2020). For instance, as Roigé *et al.* (2019) explain, ICH has been used as part of marketing strategies to promote particular destinations, to create identity brands, or to generate cultural tourism products that answer the tourist's needs for a more sustainable way of travelling and the feeling of belonging to the place they visit, even feeling like a local (UNWTO, 2019).

ICH also enables tourists to gain first-hand experience of a culture, since it includes performance activities such as festive events, rituals, social activities or oral expressions (UNESCO, 2018). These activities are often staged and performed in public spaces of cities or villages on specific dates. In these cases, the performative representation and celebration of the local culture allows even visitors to spontaneously feel and, indeed, be part of it (Chen *et al.*, 2020). In this way, ICH can facilitate encounters between visitors, residents, officials and professionals, who participate and co-create their particular experiences of the event. In this sense, cultural tourism has a crucial role in the transmission of this particular type of heritage. In this context, new ways of experiencing culture and heritage have emerged, and creative engagement with the culture has been gaining in popularity. According to Richards and Raymond (2000), creative tourism implies active participation in various activities, such as courses or learning experiences, which take place at the destination. These experiences and learning opportunities might involve arts and crafts, languages, spirituality and gastronomy, among others.

Therefore, the link between creative tourism and ICH is evident: being part of the creation process – or, in this case, we can say also staging or performing – enhances the memorability of the experience by increasing tourists' interaction and engagement (Richards, 2020). Thus ICH is increasingly becoming an attraction for tourism, since it offers a common language that allows visitors to participate and gain a sense of integration into the culture of a destination. In this sense, ICH experiences connect with tourism's quintessence, i.e. to be exposed to, and have the option to have an encounter with, 'other' people (Robinson and Smith, 2006). Although the deployment of ICH in the tourism context is considered a way to preserve local culture, its commercialization is not free from challenges. ICH's growing role in the tourism context emerges from issues about heritage sustainability (Richards, 2018). For instance, when such commercialization is not regulated, external stakeholders (such as non-local companies) might end up leading this process (Sevilla, 2017). This has the potential to cause an unjust distribution of the resultant economic benefits, as well as threatening the transmission of the very heritage being 'protected' to the next generation, should the community come to feel that it is no longer their own. As du Cros and McKercher (2020) state, co-operation and participation of different types of people in the ICH is essential to keep this heritage alive and pass it through one generation to another.

In this context, communication, co-ordination and collaboration among stakeholders are needed to preserve and promote intangible heritage in a beneficial way (Io, 2019). Therefore, heritage marketing must balance satisfying visitors' needs and preserve and transmit cultural heritage (Misiura, 2006).

Thus tourism should preserve the economic and social values of ICH and its cultural meanings in order to use it as a sustainable resource (Casanovas and Arcos-Pumarola, 2017; Kim *et al.*, 2019).

To preserve ICH's cultural and social meanings also means to consider the way it is communicated and presented to others. That is so because heritage preservation requires a previous heritage's understanding by everyone involved, that allows appreciating its value and perceiving its preservation as worthy (Fontal Merillas, 2013; Cerquetti and Ferrara, 2018). Reliable and didactic information about ICH is, therefore, necessary to enable a sustainable approach to heritage. Taking this into account, in the present context, where the communication environment has been disrupted by digital technologies (Xue *et al.*, 2019), stakeholders involved in the promotion and management of ICH must take into account its presence in social networks.

11.2.2 The importance of social networks in promoting intangible cultural heritage

The perceived image of a destination, which is the one made up in one's mind based on the images that the destination emits (Osácar Marzal, 2016), is constructed through a dialectic process in which the tourist's cultural background interacts with multiple images projected by various agents such as DMOs, advertising, films and series. At the same time, the resultant image perceived by the tourist will feedback and take part in other dialectic processes of image construction (Marine-Roig, 2015).

To preserve and transmit the meaning and values of ICH, it is essential to pay attention to the messages emitted by the various agents. Traditionally, the projected images that are involved in the process of constructing an overall image of a destination have been divided into two categories: organic or induced (Osácar Marzal, 2016; Osácar *et al.*, 2017). Induced images are those formed by the agents that take part in the promotion of a destination, for instance DMOs. Organic images are usually not created by the purpose of advertising and not controlled by the DMO, and they are excluded from the destination's official narrative account. Multiple agents generate these organic images; for example, TV, radio, movies, literature and other popular cultural expressions are major actors in constructing a particular gaze over a particular territory and have the infrastructure to share it massively (Iwashita, 2006).

Nevertheless, in recent years, the rise of a new actor has disrupted the process of constructing a destination's image. The rising popularity of the internet has transformed how the tourist gathers information about the destination they plan to visit (Hays *et al.*, 2013). In this context, social networks have gained importance in the process of creating the image of a place. One unique aspect of these platforms is that in social networks, induced and organic images are mixed. This enables the tourist to check the official narratives and messages about a destination against the particular opinions of individual travellers. The fact that destinations do not have full control of their image is emphasized by the increasingly important role of these social networks (Marine-Roig and Huertas, 2020).

Traveller-generated content (TGC) is, therefore, a new agent of organic image formation (Marine-Roig, 2019) that has overtaken the capacity for influence of the other image-formation agents by virtue of its authenticity. This has occurred because TGC is formed from the opinions of tourists who have first-hand experience of the destination and are transmitting it to the rest of the social network's users.

In this context, the predominance of TGC in modulating tourists' desires might be threatening for heritage preservation. The popularization of certain places or ICH expressions might draw the attention of numerous travellers. The ability of DMOs to foresee and manage these trends is crucial to prevent negative consequences, such as damage to tangible elements or the banalization of ICH expressions. The case of the Coves de Can Riera in Torrelles de Llobregat, near Barcelona, illustrates the challenges that TGC poses to managing unexpected interest from tourists due to social networks. In this case, the city council of Torrelles was forced to ban access to the caves because a massive influx of visitors was damaging the location. The visitors were attracted to the caves after many pictures on Instagram and blogs disclosing a hidden gem near Barcelona. Since 2019, the city council has been examining how to regulate the visit to the caves and preserve this fragile natural phenomenon (Ajuntament de Torrelles de Llobregat, 2019).

Taking this into account, and considering the importance of preserving the cultural meanings given by each community to its ICH, analysing how this ICH is communicated and shared by the different users of social networks is essential to anticipate and preserve its authenticity in a cultural tourism context. Analysing this aspect is also essential for heritage stakeholders aware of social media opportunities to get a place in the mind of potential visitors since their strategies have to consider different aspects, including the interaction with users (Surugiu and Surugiu, 2015).

11.2.3 The case of the *Fête de l'Ours* in France

In the international context, in 2020, France could count a total of 45 heritages sites recognized by UNESCO. Thirty-nine of these are cultural heritage sites (UNESCO, 2020b), i.e. specific places with outstanding tangible heritage resources. Beyond these places associated with tangible heritage, some specific intangible cultural heritage expressions are also recognized by UNESCO. Concretely, UNESCO protects 18 intangible cultural expressions in France, including the summer solstice celebration in the Pyrenees and the techniques and knowledge related to dry stone walling (UNESCO, 2020a).

Beyond UNESCO's international seal of recognition, each country has its own designations and structures for preservation. In the case of France, and narrowing it down to intangible cultural heritage, the Inventaire National du PCI (PCI standing for the French translation of ICH, 'patrimoine, culturel, immatériel') should be considered (Ministère de la Culture, 2020a). The Culture Ministry of France manages this inventory. To apply for a UNESCO nomination, the ICH element must previously have been included in the Inventaire National du PCI (Ministère de la Culture, 2020a).

Concurrently, there is the designation Villes et Pays d'Art et d'Histoire, which consists of a network that gathers 196 territories, cities and villages around France. Under this designation, the places are committed to promoting and spreading the value of architecture and tangible and intangible heritage. In this sense, they offer educational activities, guided visits and exhibitions aimed at all public members: citizens, visitors and young people. According to Ministère de la Culture (2020b), to be included in this network the locale must fulfil four criteria. First, there must be a demonstrable political commitment to ground its development on culture, architecture and heritage. Second, the village, city or territory must have internal consistency; that is, it has to be historically, geographically, demographically and culturally coherent. Third, the community must guarantee the required infrastructure is in place to ensure knowledge, preservation and improvement of the landscape, architecture and heritage. Finally, the stipulations of the Villes et Pays d'Art et d'Histoire project must be applied to all municipality policies affecting the designated area.

The case study we present is located in Occitan region, in an area known as Val du Tet et du Tech, a member of the Villes et Pays d'Art et d'Histoire network (Ministère de la Culture, 2020b). In this sense, this case study is framed in a policy and cultural context that ensures its recognition and preservation and, at the same time, fosters the valorization of and community involvement with the analysed expression of ICH. The case study focuses on the *Fête de l'Ours*, a festival in which a representation of a bear hunt takes place. A man dressed as a bear walks down from the mountains and is chased by the citizens through the streets of the village. This representation takes place in three villages in Haut Vallespir, a region located in the south of France near the border with Catalonia (Fig. 11.1). These villages are: Arles-sur-Tech, Prats-de-Molló-la-Preste and Saint-Laurent-de-Cerdans.

The origins of the festival are linked to oral traditions and the way the three villages perform the festival share some similarities. Some documents date these origins to the 9th century. So this is considered as an ancient festivity, for which we have written evidence beginning from 1835. The first written description of the festival appears in the book *Histoire de Roussillon: comprenant l'histoire du Royaume de Majorque* (Pays Pyrénées-Méditerranée, 2014). Although nowadays the *Fête de l'Ours* occurs in only three villages, it once took place in more of the villages of Haut Vallespir. In the case of Prats-de-Molló-la-Preste, Arles-sur-Tech and Saint-Laurent-de-Cerdans, the festival underwent a revival at the end of the 20th century when several ethnographers became interested in it (Pays Pyrénées-Méditerranée, 2014).

The number of visitors and participants in the *Fête de l'Ours* has grown since it appeared in various documentaries. Among these, it is worth mentioning the one filmed by Chegaray and Breugnot in 1979, which was broadcast on French public television (Pays Pyrénées-Méditerranée, 2014).

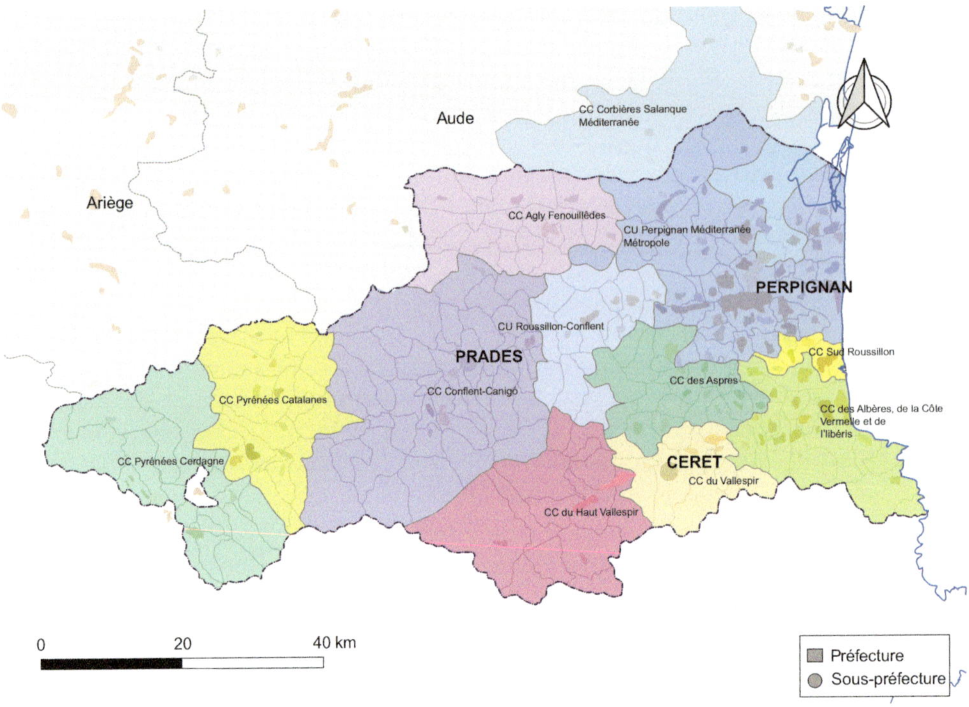

Fig. 11.1. Department of Pyrénées-Orientales (location of Haut Vallespir).

From that moment on, the *Fête de l'Ours* became more popular and generated proximity tourism, both from Catalonia and France. Although the festivity has familiar features, some variations can be identified. Mainly, we can distinguish the bear costume, the roles of people staging the festivity, and the weekend in which it takes place. The last of these may correspond to the aim of enlarging the festivity in the region in order to encourage the regular influx of visitors and manage the flow of tourists in the best possible way.

Looking in more depth at how the festival proceeds, we can say that there are three stages, over which many different actors appear. In the first stage, the young man who plays the role of the bear, wearing a costume made of fur,[1] walks down from the mountains. In a second stage, the bear invades the village and starts to hunt and kidnap people, especially those dressed as young girls. This chase happens along the streets and squares. Also in this stage, men dressed as hunters appear. They chase the bear and try to prevent the kidnappings. During this chase, the bear may touch people's faces and leave marks or 'footprints' of soot. This is the moment when the performance is at its most participative and integrative, since given the increasing popularity of the festivity and the growing participation of visitors, being attacked by the bear or being invited to dance with it has become a symbol of inclusion in the festivity. Finally, the performance ends with the bear being chased in the main village square. Men dressed all in white, representing barbers, descend upon the bear and shave it. This public shaving symbolizes the reintroduction of the bear, who is now a man, into society.

One guiding thread of the festival is that the bear's walk down from the mountains, the hunt through the village streets and its shaving are accompanied with traditional songs and dances, such as the one known by the name *Ball de l'Ós*. Music is played by a group of musicians with traditional Catalan instruments. A marching band plays traditional music when the bear is chased through the streets. Then, an ensemble known as a *cobla* – group of 11 musicians playing wind instruments – accompanies a dance called the *Ball de l'Ós* when it is shaved. In all three villages, the *Fête de l'Ours* ends with the popular *sardana* dance, which is also typically played by the *cobla* (Villes & Pays d'Art & Histoire, 2019).

The *Fête de l'Ours* represents a tradition that has evolved with society and its needs. For example, formerly this celebration took place on 2 February, symbolically marking the end of the winter season when bears leave hibernation. Based on the behaviour of the local bears, which are today no longer found in the region, the villagers knew whether spring would soon come or winter would last longer. Nowadays, however, the celebration has been rescheduled to fit French school holidays, and it is celebrated during the first, second and third weekends of February. In Saint-Laurent-de-Cerdans, it is celebrated during the third weekend of February, running in parallel with the village's annual carnival.

The villages which still celebrate it have recently come together to consolidate their activities and promote the relevance of the festival. In this vein, since 2014, Arles-sur-Tech, Prats-de-Mollo-La-Preste and Saint-Laurent-de-Cerdans have been working together to prepare a joint application for UNESCO intangible cultural heritage status (Villes & Pays d'Art & Histoire, 2018). Their activities have been conducted under the territorial designation Pays Pyrénées Mediterranée, in order to achieve worldwide recognition. This highlights the fact that the values that stand out from this festival and which validate its candidacy for ICH status are not only its substantial cultural value, which is deeply linked to the identity of its people and the region, but also a sense of territorial cohesion. With the documentation prepared, the *Fête de l'Ours* was automatically included in the French *Inventaire National*; meanwhile UNESCO is still deliberating their candidacy.

11.3 Methodology

In order to achieve the objectives of the study, a quantitative methodology was established. The research focused on one particular social network: Instagram. Instagram is a social network on which people can share pictures and opinions about their trips, places they have eaten, places they have slept or activities they have participated in. In this sense, Instagram, being primarily an image-based resource, has become one of the most popular platforms among social media users. In its ten years of existence, the network has grown from approximately one million active users per month in 2010 to around one billion per month in 2020, becoming the sixth most widely used social network (We Are Social and Hootsuite, 2020, p. 123). It is clear, then, that Instagram is a powerful tool to share and sell products all over the world. Seeing its potential in this regard, tourist destinations are using this platform to promote their attractions and heritage. Taking this context as a framework, Instagram was selected as the primary source of information for this case study.

To identify pictures to be analysed, a search was conducted on Instagram for hashtags likely to be used by users posting images of the *Fête de l'Ours*. These hashtags were written in both Catalan and French, since the *Fête de l'Ours* is celebrated near the Catalan border and therefore has strong links to Catalan territory. This primary search, conducted on 7 September 2020, showed that most people use the French hashtag #fetedelours (861 times), as shown in Table 11.1. It is also worth mentioning that out of the 1319 hashtags naming the festival that were used in the analysed posts (bearing in mind that many users added multiple hashtags), 1012 (77%) were variants of the French name for the festival, compared to 307 (33%) that were variants of the Catalan name (*Festa de l'Ós*). This informs us that even though the festivity is located in a border territory, it has a broader impact in French culture.

The search period was restricted to posts from 2013 onwards, since the images available on Instagram started from that year. That meant that the sample to be analysed consisted of 571 posts. From these, 293 were discarded – reducing the number of posts to 278 – because the images did not show the *Fête de l'Ours* of Haut Vallespir region, but were pictures taken during an event related to bears celebrated annually since 2014 in Paris. This event might be related to the year the *Fête de l'Ours* was included in the *Inventaire National du PCI*. This sample was refined again, accepting only those pictures whose author could be identified as a visitor, local, official/institutional or professional. In the end, 137 images were retained and formed the final sample for analysis.

In order to gather information to achieve the research objectives, an *ad hoc* tool was created to compile the information, in the form of an observation table composed of three sections. The first section detailed the type of image posted, the second classified the profile type of the user who posted the picture of the festival, and the third identified the type of written message (if any) posted with the picture.

Table 11.1. Analysis of Instagram hashtags of the festival's name, conducted in September 2020.

Hashtag	Number of hashtag appearances
#festadelos	207
#festadelós	100
#fêtedelours	151
#fetedelours	861

Regarding the type of image, the pictures were classified into five categories:

- 'during the fête': images that show the bear walking down towards the village, the bear dirtying people's faces, the public shaving, etc.;
- 'preparing the fête': images that show the players of the main roles putting on their costumes, doing their make-up and so on;
- 'festive atmosphere': images showing people drinking beer and other beverages, etc.;
- 'specific detail': images showing, for example, the bear's costume, the bear's footprints, or other details of the festival's features;
- 'others': images that cannot be classified into the categories above.

Users' profiles were also classified into four categories:

- 'visitor': if the profile said the user was not from Arles-sur-le Tech, Prats-de-Mollo or Saint-Laurent-de-Cerdans;
- 'locals': if the profile said the user was from one of these three villages;
- 'official/institutional': if the profile is the official account of a regional body, a public institution or heritage centre carrying out activities in the region, a TV or radio channel, or a celebrity;
- 'professional': if the user specifies that they are, for instance, a photographer, a make-up artist – someone with a professional connection to the festival.

A fifth category, 'does not say', classified profiles where this information was not provided. However, as mentioned above, such pictures were excluded from the final sample, which included only those whose author could be classified into one of the categories mentioned above.

Finally, the type of message attached to the image was classified into three categories:

- 'descriptive information': when the user only wrote the festival's name in the image description, or gave only brief descriptive information, for example about what is happening in the picture, the dates of the festival or some other information;
- 'explaining the experience': for photographs where the user described what the festivity is about, or how they felt about it;
- 'there is no message': for when the user only wrote hashtags in the image description field.

11.4 Results

In this section, the results obtained from the content analysis are presented. This analysis is divided into three sub-sections. The first section is concerned with the type of images posted; the second with the users' profiles; and the third with the content of the textual messages that accompany the image.

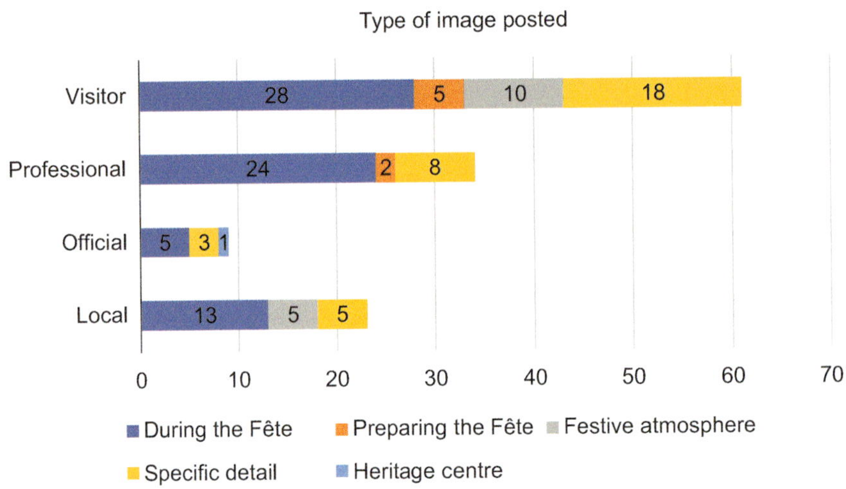

Fig. 11.2. Type of image posted.

11.4.1 Type of image

When considering the type of image posted by each identified profile, it can be seen that in all profile types, it is most common to share pictures of activities that take place during the *Fête* (Fig. 11.2). In total, 70 pictures fitted this category, representing the dramatic aspects of the festival as they were performed.

The second most frequently shared type of image, with a total of 34 posts, is photos that focus on specific details of the celebration, that is sooty faces, the various costumes used, etc. In these pictures, the particular detail was extracted from the broader context of the staging and valorized through the photograph. These images highlight how relevant tangible elements are in the context of intangible celebrations, making it clear how tangible and intangible heritage are closely linked.

The third most frequent type of image shows people enjoying the atmosphere of the festival. Fifteen posts were classified under this category. This group included images of people having fun but not involved in staging the tradition. It is interesting to note here that only visitors and locals shared this type of image: official institutions restrict their interpretation of the *Fête* to the staging of the tradition and do not share images that show the more casual, informal aspects of the event – for example of people having fun together on the periphery. The preparations for the festival, with eight posts, attracts the attention of both visitors and professional photographers, but official institutions and locals do not valorise this part of the celebration on Instagram.

This analysis allows us to affirm that the traditional staging and particular tangible elements of the celebration are its main attractions, according to the number of pictures posted by users of all categories. However, the visitors' gaze over the *Fête de l'Ours* – and specifically their interest in other elements, such as preparations for the drama – suggests that the promotion and account of this celebration by official institutions on social networks could be widened by showing, for example, how the role-players prepare for their performance.

11.4.2 Users' profiles

Users were classified into four different types according to their profiles. By far the greatest number of posts about the *Fête de l'Ours* – 64 out of 137, or 47% – were from visitors, as shown in Fig. 11.3.. The vast majority of these were proximity visitors; that is, they came from other parts of France or Catalonia. However, it should be taken into account here that when the sample was formed, many posters of *Fête de l'Ours* images were excluded from the sample, which is not taken into account for this analysis, because the authors did not have enough information about their location to determine whether they were locals or visitors. While those who run official and professional accounts usually give their location in their profile, some individual users choose to withhold this information. As such, our result of 47% of posts by visitors may under-represent the true proportion of visitors posting photographs of the *Fête de l'Ours* on Instagram.

In second place, professionals who focused on the *Fête de l'Ours* and shared their work on Instagram were identified. Nineteen per cent of profiles included in the analysis belonged to local people who live in the towns where the *Fête de l'Ours* is celebrated. For the same reasons described above, across all the posts on the festival on Instagram, this proportion could be greater than what the findings show, since some inhabitants of

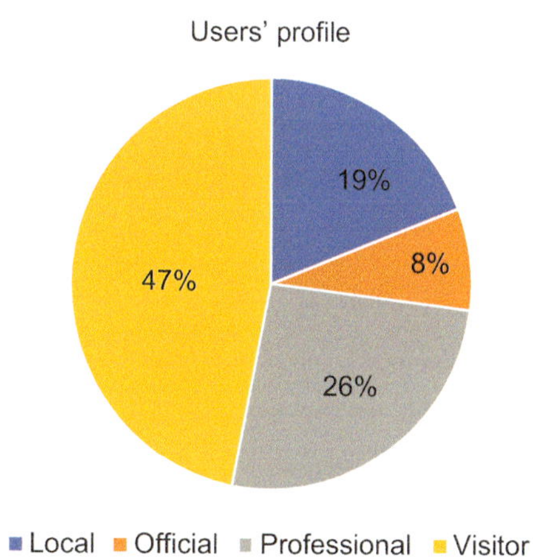

Fig. 11.3. Users' profile.

these towns who posted images may have avoided sharing information about where they live. Finally, only 8% of posts were published by official/ institutional users.

11.4.3 Type of message

As shown in Fig. 11.4., most of the posts (70 out of 137) only included brief descriptive information, for example, about what is happening in the picture, the dates of the festival or some other information.

A further 51 posts only included hashtags related to the content of the photograph. Nevertheless, it is interesting to note that Instagram enables users to share more complex messages that explain the development of the staging and how the festivity is experienced. Under the category 'Explaining the experience', 15 posts were included which not only offer descriptive information but also explain the meanings and the feelings the poster experienced during the various parts of the event.

This analysis shows that posts which expand the narrative and take advantage of the social network's possibilities are in a minority among all user types. This applies even for official profiles: in our sample, only one post from an official profile fitted this category.

11.5 Discussion and Conclusions

The present study examined how the *Fête de l'Ours* is being presented on the Instagram social network by users of various different profile types. This analysis has revealed which particular elements or moments of the festivity are highlighted by locals, visitors, professionals and official/institutional

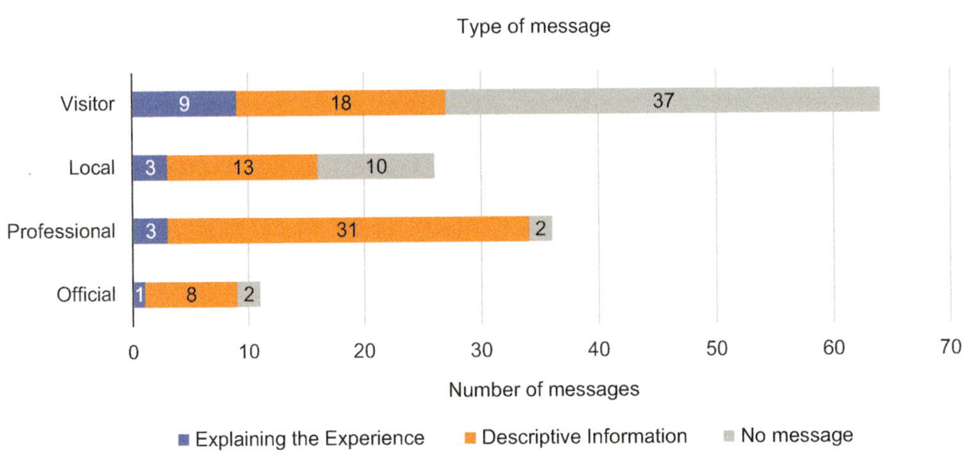

Fig. 11.4. Type of message.

channels, and has also highlighted the different ways in which these users interact with the social networking tools provided by Instagram.

In a context where the image of a destination is not only dependent on the official narrative (Marine-Roig and Huertas, 2020), the study contributes to the existing literature by explicitly stating that visitors are the most proactive agents in posting about and explaining the analysed festival. This kind of traveller-generated content (TGC) is not only the most influential agent in the formation of the image of a particular destination or tourist attraction (Marine-Roig, 2019), but has also become the most prolific agent in the image-formation process by virtue of its continuous growth. This is the case because the number of locals and institutions remains roughly the same, while visitor numbers increase cumulatively over time, as new people are attracted each year. In the context of the case study, new visitors who have not previously experienced the *Fête de l'Ours* will probably post more images than local residents who see it every year, thus feeding a growing collection of TGC. In this context, the role of official institutions should distinguish their voice in a social media context in order to foster the balance between satisfying visitors and preservation (Misiura, 2006). Thus, the capacity of local stakeholders to provide the visitor with reliable and trustworthy information is vital for creating a later TGC that the local community is comfortable with. This way, the balance should be found between posting interesting and appealing information to visitors and that which is respectful of the meaning that ICH has for the local community.

The content analysis of the posts revealed that while visitors are sharing the whole process of the festival, including its preparation and the festive atmosphere surrounding it, the official Instagram accounts tend to focus more on the main attraction of the *Fête* itself, namely the drama of the bear hunt. However, most of the posts offer only brief descriptive information on the text section. These results suggest that there is room for official accounts to improve their influence on social networks, not only by posting more often but also by expanding the way they show and share the *Fête de l'Ours*.

The analysis found that official accounts narrowly focused on the dramatic aspects of the festival. Therefore, they were not offering information about the meanings of the festival, or the characters that populate it, leaving individual visitors to fill this gap. The generation of high-quality content could be used to distinguish official information from that generated by other users and, in this way, enhance the influence of official agents in the image-formation process. For example, to attract and hold the attention of users, official bodies could go beyond using merely descriptive information in their posts and include more images of different aspects of the festival, and enhance their posts with more complex and informative texts, providing information about the tradition, sharing anecdotes, telling stories, or seeking interaction from other users. In the end, the aim is to communicate a relevant and 'unique experience'(du Cros and McKercher, 2020). These actions could help spread knowledge of the meanings of the *Fête* among the Instagram-using public. This is especially important in an ICH context, since the role of these agents is not only to promote the festivity, but also to preserve and share its cultural meaning and values. Thus, a good recommendation for official profiles could be to take a more proactive role in creating appealing and didactic messages about the *Fête de l'Ours* in the context of social networks. The didactic approach could differentiate official profiles discourse in social networks, placing them as reliable sources of information and promoting heritage's understanding and, therefore, its preservation in the long term (Fontal Merillas, 2013; Cerquetti and Ferrara, 2018). To this end, co-ordination among the different stakeholders involved in a particular ICH could amplify their social networks presence (Io, 2019). Co-ordination also helps build a coherent and shared narrative of ICH among all stakeholders, an essential trait for impacting the social media context.

To gain more visibility in a context where TGC is quantitatively more prominent, official institutions should also consider publishing posts about their ICH events during the whole year and not only during the celebration. Summarizing these findings, Table 11.2 offers a comprehensive perspective of the improvement opportunities identified after the analysis. In this sense, Table 11.2 relates the findings explained above with the recommended actions for the stakeholders.

These suggestions could help support the candidacy of the *Fête de l'Ours* for UNESCO ICH status by demonstrating the will of the local official bodies to spread the meaning of the festival and enhance awareness of it among other communities. The role of official Instagram profiles in this context is also crucial to avoid folklorization of the *Fête*; in this sense, their proactivity in creating a complex and meaningful message is essential to generate content

Table 11.2. Findings of the research and possible actions.

Findings	Recommended actions for stakeholders
Official Instagram accounts tend to focus on the main attraction of the drama of the bear hunt.	Offer richer contextual information about the meanings of the festival. For instance, other posts related to the preparation of the festivity and anecdotes can be made, so visitors get the sense and the meanings behind it.
Posts offer only brief descriptive information.	Creation of high-quality content including more extensive and didactic explanations. Use of engaging information (sharing anecdotes, telling stories…) that seeks interaction from users.
Most of the posts are published by visitors during their experience.	Official accounts should post photographs and make 'stories' regularly to have recent publications during the whole year.
The different official Instagram accounts publish a few posts with a brief description and do not transmit a shared message between the different institutions involved.	Foster collaboration and co-ordination between the different stakeholders in social networks to amplify their social network presence and build a coherent narrative account of the ICH.

that shares the values and identity of this celebration and its connections to the territory.

Regarding the limitations, this research was based on one social media platform: Instagram. Further research could explore this festivity in other social networks such as Twitter and Facebook or even TikTok, since each social network has its tools to create content and have different audiences. It would also be interesting to run interviews and surveys to citizens and visitors to evaluate the festivity's feelings. Another limitation of this investigation is that it is only applied to one ICH in France. Future research could compare the narrative account in social networks of different ICH cases that UNESCO has already recognized with others that are not. Finally, a set of indicators could be created to measure the degree of involvement that an ICH can set and the feelings that the involvement generates between citizens and visitors after its performance.

Note

[1] Only in the town of Saint-Laurent-de-Cerdans does the bear wear a costume made from actual bear fur, imported from Canada.

Acknowledgements

This work was supported by the project *Patrimonio Inmaterial y Políticas Culturales: desafíos sociales, políticos y museológicos* (PGC2018-096190-B-I00), funded by the Spanish Ministry of Science, Innovation and Universities / Spanish State Research Agency / European Regional Development Fund.

References

Ajuntament de Torrelles de Llobregat (2019) L'Ajuntament de Torrelles prohibeix l'accés a les coves de Can Riera. Available at: http://www.torrelles.cat/actualitat/noticies/2019/04/15/l-ajuntament-prohibeix-l-acces-a-les-coves-de-can-riera (accessed 12 October 2020).

Casanovas, O. and Arcos-Pumarola, J. (2017) Protección del patrimonio y su contexto: de la conservación a la preservación. In: Coma, L. and Santacana, J. (eds) *Ciudad Educadora y Turismo Responsable*. Trea, Gijón, Spain, pp. 213–232.

Cerquetti, M. and Ferrara, C. (2018) Marketing research for cultural heritage conservation and sustainability: lessons from the field. *Sustainability* 10(3), 1–16. DOI: 10.3390/su10030774.

Chen, Z., Suntikul, W. and King, B. (2020) Constructing an intangible cultural heritage experiencescape: the case of the Feast of the Drunken Dragon (Macau). *Tourism Management Perspectives* 34(4), 1–11. DOI: 10.1016/j.tmp.2020.100659.

du Cros, H. and McKercher, B. (2020) *Cultural Tourism*, 3rd edn. Routledge, London.

Fontal Merillas, O. (2013) *La educación patrimonial: del patrimonio a las personas*. Trea, Gijón, Spain.

Hays, S., Page, S.J. and Buhalis, D. (2013) Social media as a destination marketing tool: its use by national tourism organisations. *Current Issues in Tourism* 16(3), 211–239. DOI: 10.1080/13683500.2012.662215.

Io, M.U. (2019) Collaboration between practitioners and public agencies in preserving and promoting musical heritage in Macao. *Journal of Heritage Tourism* 14(1), 19–32. DOI: 10.1080/1743873X.2018.1452248.

Iwashita, C. (2006) Media representation of the UK as a destination for Japanese tourists: popular culture and tourism. *Tourist Studies* 6, 59–77.

Kim, S., Whitford, M. and Arcodia, C. (2019) Development of intangible cultural heritage as a sustainable tourism resource: the intangible cultural heritage practitioners' perspectives. *Journal of Heritage Tourism* 14(5-6), 422–435. DOI: 10.1080/1743873X.2018.1561703.

Marine-Roig, E. (2015) Identity and authenticity in destination image construction. *Anatolia* 26(4), 574–587. DOI: 10.1080/13032917.2015.1040814.

Marine-Roig, E. (2019) Destination image analytics through traveller-generated content. *Sustainability* 11(12), 1–23. DOI: 10.3390/su11123392.

Marine-Roig, E. and Huertas, A. (2020) How safety affects destination image projected through online travel reviews. *Journal of Destination Marketing and Management* 18(3), 1–11. DOI: 10.1016/j.jdmm.2020.100469.

McKercher, B. (2020) Cultural tourism market: a perspective paper. *Tourism Review* 75(1), 126–129. DOI: 10.1108/TR-03-2019-0096.

Ministère de la Culture (2020a) Patrimoine Culturel Immatériel. Patrimoine Culturel Immatériel Available at: https://www.culture.gouv.fr/Sites-thematiques/Patrimoine-culturel-immateriel/L-inventaire-national-du-PCI (accessed 1 October 2020).

Ministère de la Culture (2020b) Villes et Pays d'Art et Histoire Available at: https://www.culture.gouv.fr/Aides-demarches/Protections-labels-et-appellations/Label-Ville-et-Pays-d-art-et-d-histoire#attribution (accessed 2 October 2020).

Misiura, S. (2006) *Heritage Marketing*. Butterworth-Heinemann Heritage Marketing, Oxford, UK.

Narangajavana Kaosiri, Y., Callarisa Fiol, L.J., Moliner Tena, M.Á., Rodríguez Artola, R.M. and Sánchez García, J. (2019) User-generated content sources in social media: a new approach to explore tourist satisfaction. *Journal of Travel Research* 58(2), 253–265. DOI: 10.1177/0047287517746014.

Osácar, E., Font, J. and Colom, M. (2017) La imagen turística y la identidad de un destino. In: Coma, L. and Santacana, J. (eds) *Ciudad Educadora y Turismo Responsable*. Trea, Gijón, Spain, pp. 179–193.

Osácar Marzal, E. (2016) La imagen turística de Barcelona a través de las películas internacionales. *PASOS Revista de Turismo y Patrimonio Cultural* 14(4), 843–858. DOI: 10.25145/j.pasos.2016.14.055.

Pays Pyrénées-Méditerranée (2014) 'Les fêtes de l'Ours du Haut-Vallespir: Prats-de-Mollo-la-Preste'. Ministère de la Culture. Prats-de-Mollo-La-Preste.

Richards, G. (2018) Cultural tourism: a review of recent research and trends. *Journal of Hospitality and Tourism Management* 36(2), 12–21. DOI: 10.1016/j.jhtm.2018.03.005.

Richards, G. (2020) Designing creative places: the role of creative tourism. *Annals of Tourism Research* 85(1), 1–11. DOI: 10.1016/j.annals.2020.102922.

Richards, G. and Raymond, C. (2000) Creative tourism. *Atlas* 23, 16–20.

Robinson, M. and Smith, M. (2006) Politics, power and play: the shifting contexts of cultural tourism. In: Robinson, M. and Smith, M. (eds) *Cultural Tourism in a Changing World: Politics, Participation and (Re)presentation*. Channel View Publications, Clevedon, UK, pp. 1–17.

Roigé, X., Mármol, Cdel., Guil, M. and Universitat de Barcelona (España)., (2019) Los usos del patrimonio inmaterial en la promoción del turismo. El caso del Pirineo catalán. *PASOS, Revista de Turismo y Patrimonio Cultural* 17(6), 1113–1126. DOI: 10.25145/j.pasos.2019.17.077.

Ruiz, I. (2020) Marketing turístico y fiestas locales: estudio de caso de las Fallas de Valencia. *Cuadernos de Turismo* 45, 363–380.

Salvador, M. and Abellan, N. (2020) Las tejedoras mayas de Guatemala: un proceso activo para la salvaguardia de su patrimonio cultural inmaterial. *Tourism & Heritage Journal* 2, 93–109. DOI: 10.1344/THJ.2020.2.7.

Sevilla, A. (2017) Los riesgos de la patrimonialización: una reflexión. In: Pérez, M.L. and Machuca, A. (eds) *La patrimonialización ¿un nuevo paradigma?* Colegio de Etnólogos y Antropólogos Sociales A.C., Ciudad de México, pp. 27–32.

Su, X., Li, X. and Kang, Y. (2019) A bibliometric analysis of research on intangible cultural heritage using CiteSpace. *Sage Open* 9(2), 1–18. DOI: 10.1177/2158244019840119.

Surugiu, M.-R. and Surugiu, C. (2015) Heritage tourism entrepreneurship and social media: opportunities and challenges. *Procedia – Social and Behavioral Sciences* 188(2010), 74–81. DOI: 10.1016/j.sbspro.2015.03.340.

UNESCO (2018) Basic texts of the 2003 Convention for the Safeguarding of the Intangible Cultural Heritage. Available at: https://ich.unesco.org (accessed 1 October 2020).

UNESCO (2020a) List of Intangible Cultural Heritage. Available at: https://ich.unesco.org/en/lists?text=&country[]=00077&multinational=3&display1=inscriptionID#tabs (accessed 1 October 2020).

UNESCO (2020b) Patrimoine Mundiel France. Available at: https://whc.unesco.org/fr/etatsparties/fr (accessed 1 October 2020).

UNWTO (2019) International Tourism Highlights. 2019 edition. UNWTO. Available at: https://doi.org/10.18111/9789284421152 (accessed 1 October 2020).

Vecco, M. (2010) A definition of cultural heritage: from the tangible to the intangible. *Journal of Cultural Heritage* 11(3), 321–324. DOI: 10.1016/j.culher.2010.01.006.

Villes & Pays d'Art & Histoire (2018) El Patrimoni Cultural Immaterial a les Valls Catalanes. Available at: https://www.valleescatalanes.org/ca/el-patrimoni-cultural-immaterial-les-valls-catalanes (accessed 27 August 2020).

Villes & Pays d'Art & Histoire (2019) Focus fêtes de l'ours haut Vallespir. Available at: https://www.valleescatalanes.org/sites/default/files/Images/final_focus_ours_francais_0.pdf (accessed 13 September 2020).

We Are Social and Hootsuite (2020) Digital 2020: global digital overview. Available at: https://datareportal.com/reports/digital-2020-global-digital-overview (accessed 1 October 2020).

Xue, K., Li, Y. and Meng, X. (2019) An evaluation model to assess the communication effects of intangible cultural heritage. *Journal of Cultural Heritage* 40(2), 124–132. DOI: 10.1016/j.culher.2019.05.021.

Conclusion

Dimitrios Stylidis[1]* and Nikolaos Boukas[2]

[1]*Middlesex University London, UK*
[2]*European University Cyprus, Cyprus*

This edited volume focused on tourism marketing in western Europe, exploring issues related to tourism product development, market segmentation, destination image, promotion and positioning in this sub-region of the world. The book included 11 cases of the following European countries/areas: Alps, Austria, Italy, France, Germany, Greece, Greenland, Portugal, Spain and the UK. The book adds to current knowledge in tourism marketing theory and practice by discussing the challenges and issues tourism destination marketers are facing in this part of the world.

The first chapter by Mengyun Hu, Eleonora Pantano and Nikolaos Stylos discussed how the internet of things (IoT) affects the travel experience. Their contribution offered a thorough presentation of the applications of the IoT in destination management (e.g. attractions, accessibility), assisting DMOs to link to travel markets, along with understanding tourists' preferences and behaviours. Next, the role of the IoT in hospitality was presented (e.g. self-check-in and -check-out, room controlling), followed by its applications in transportation (e.g. real-time traffic control, smart airports) and in travel intermediaries (compare and book tickets, accommodation, rental cars). The potential of the IoT within the tourism context for informing, contextualizing, personalizing, remote controlling and communicating, have a tremendous effect on visitors' experiences throughout the timespan of a trip. The chapter concluded by proposing a great array of recommendations for academics and practitioners alike such as contactless technologies to minimize the impacts of Covid-19 on tourists' decision-making, perceived value and satisfaction. Overall, the study serves as an excellent starting point for further exploring various issues and applications of the IoT on destination management, hospitality, transportation and travel intermediaries.

The second chapter, authored by Ilma Aulia Zaim, empirically explored the way young British tourists seek tourism-related information sources. The results of 25 in-depth interviews conducted with British internet users aged 18–34 showcased a great shift in the various types of information sources young British people use. There were five primary sources identified including: (i) suggested links by Google (Wikipedia, blogs); (ii) YouTube; (iii)

*Corresponding author: d.stylidis@mdx.ac.uk

 Tourism Marketing in Western Europe (ed. N. Boukas and D. Stylidis)

DOI: 10.1079/9781789248753.conc

social networking sites (Instagram, Facebook, Snapchat); (iv) personal sources of information (previous visits, recommendations by friends/relatives); and (v) travel guidebooks. Google appears to serve as the first starting point, whereby young British search for destination images, facts etc. YouTube videos offer more direct/realistic views of the destination through amateur travel vlogs. Social networking sites including Facebook and Instagram were mentioned as core sources of information for participants, although acknowledging that visual material is often edited by DMOs to increase destinations' attractiveness. Traditional word-of-mouth remains a credible source of information, while travel guidebooks including Lonely Planet and Eyewitness Travel offer travel recommendations, especially when young British have already decided to visit the destination. Overall, the study highlighted the importance of source credibility for British tourists while seeking travel-related information.

The third chapter, by Bernardo Borges and Rui Augusto Costa, examined local destination management organizations and the tourism marketing channels they use to promote destinations, focusing on the Centre of Portugal region. The chapter, in particular, aimed to assess DMOs' digital marketing strategies to understand their impact on tourists searching for information in the pre-visit stage, and also during the actual visit. Data were collected via 144 questionnaires distributed via tourist information offices and completed by visitors in central Portugal. One third of the visitors were domestic tourists and the rest were international, mainly from France and Spain. The results suggested that visitors in this area of Portugal used the internet, including websites and social media platforms, while searching for destination-related information in the pre-travel phase, whereas there was a greater preference for tourist offices while at the destination. The study assists local DMOs in designing and disseminating information via digital media platforms and websites that aim to interact, communicate and respond to tourists' questions especially during the pre-trip stage.

The fourth chapter by Francesco Calza, Michele Simoni, Annarita Sorrentino and Mariapina Trunfio explored a participatory planning and marketing process in demand-driven destinations, using evidence from southern Italy. Four thematic meetings conducted with 50 local stakeholders (representing hotels, camps, restaurants, bed & breakfast units, tourist agencies, policymakers and local associations) in Centola-Palinuro, a seaside destination located in the Campania region, allowed triggering of the participatory planning process. The meetings were developed under the following themes: (i) tourist experiences; (ii) quality management; (iii) accessibility, hospitality and information; and (iv) innovation and training. The increased stakeholder participation in these meetings validated the proposed theoretical framework of stakeholder engagement. Destination governance that orchestrates and facilitates sharing between the actors was assessed as fundamental in this process. The need for specific goals and strategies emerged to achieve decongestion of the much-visited coast and the alleviation of seasonality by designing new tourism experiences in resources around the countryside. Also, new tourism products such as sports activities, slow tourism, cultural short breaks and relaxation/detox were proposed to diversify the current product offering. The chapter particularly contributes to tourism marketing in the post-pandemic era, serving as a starting point in reformulating a destination's offer, especially in the post-Covid-19 crisis, to include mindfulness, digital detox and nature-based vacations.

Another niche market, that of gastronomic tourism, is the topic of the fifth chapter developed by Jesús Barreal Pernas and Gil Jannes, who focus on regions producing Protected Designations of Origin (PDO) products, and explore the similarities and differences among the profiles of international tourists visiting such PDO regions. Using the case of Spain and two well-known PDO products, olive oil and wine, the study used data collected by the Spanish National Statistics Institute recording international tourists' information preferences via survey questionnaires. Pernas and Jannes' segmentation approach based on tourists' country of origin, demographic and trip characteristics, and motivation and interests, identified four clusters with very similar characteristics in all visited regions. Notable is that three out of four visitors to wine and/or olive oil PDOs stated that they did not plan nor participated in gastronomic or rural activities. Such findings indicate that further marketing actions are required, in the context of gastrotourism, for wine and olive oil thematic tourism to develop, and to diversify Spain's tourism offering, which is currently heavily based on sea and sun. The study provides key managerial implications for further development of tourist activities related to wine and olive oil.

The sixth chapter, by Thomas Bausch, reflected on the challenges and lessons in managing hybrid destinations using the case of the Alps, shared by France, Italy, Germany, Austria, Slovenia, Switzerland and Liechtenstein. As a mountainous destination, the Alps offer a hybrid tourism product: winter sports and recreation dominate in the winter, while hiking, biking and climbing are key tourist activities in the summer. The study identified three types of destination hybridity: uniform non-hybrid, binary-hybrid and multiple-hybrid. Such hybrid products are in need of destination marketing activities that focus on the diverse seasonal products and their unique target groups. An analysis of the data from the yearly-held German travel analysis Reiseanalyse revealed that Germans' destination image of the Alps varies seasonally due to the diverse tourism product offering over the year; the promotional and marketing communication tools need to address diverse tourist groups; and an increasing polarization between low-cost and high-priced luxury travelling was noted. The chapter contributes to the management and marketing of multiple-hybrid destinations in designing strategies that aim to satisfy the expectations of a range of visitors, from the typical mass-tourist to multiple-hybrid guests.

In Chapter 7, Nikolaos Boukas explored rural tourism development in Kalavrita, Greece. While Greece is a popular destination for summer vacations, its physiognomy comprises continental places and many mountains. Nonetheless, these places have not been equally treated by tourism marketing as significant destinations and for this reason, they remain underutilized. The aim of this chapter was to examine the case of the Kalavrita mountain region, located on Northern Peloponnesus, Greece, and the marketing efforts that national and regional tourism authorities undertake to foster rural tourism development. Based on qualitative research with local policymakers (representing the overall local policy as well as tourism in the region), community officials, as well as officials on behalf of the tourism sector in the area, the chapter emphasized the tourism products that the destination offers, highlighting rural tourism as a macro-niche that is worthy to be considered as a unique form of tourism. The chapter identified the main markets of Kalavrita, signified the key stakeholders for rural tourism development, and examined the challenges and opportunities of this form of tourism, as well as the necessity for the diversification of the existing tourism product offering. The study concluded with marketing implications regarding rural tourism development in the area, such as the incorporation of key stakeholders in the formation of the rural tourism experience through effective marketing communications, the creation of more functional rural products that would give the opportunity to the visitors interact better with the destination's rural features, and the establishment of a stronger brand identity for the place as a rural destination.

Chapter 8, authored by Jacey Choe, examined US tourists' food neophobia and their local food experiences in two European countries – France and Italy. Based on 420 online questionnaires collected in 2016, the study explored whether US visitors have various levels of food neophobia and tested the relationships among different food neophobia groups and their local food consumption values and behavioural intention to return for food tourism to the destinations studied. Based on the findings, Choe classified respondents into three groups representing low food neophobia (neophilics), medium food neophobia (neutrals) and high food neophobia (neophobics). Results suggested that males tend to be more neophobic than females, while married people were also more neophobic than singles. Another notable finding was that both younger and older US tourists were more neophobic than the middle-aged ones. No differences, however, were observed among respondents with different education and income levels. The study contributes to the tourism literature by confirming the presence of various neophobia sub-groups, and by establishing the negative effect the fear of novel food is having on tourists' local food experiences in a destination. The study also provides practical implications useful to French and Italian destination marketers to attract and retain US tourists.

In the ninth chapter, Elitza Iordanova focused on the challenges involved in marketing destinations linked to dark moments in history. The study, in particular, examined how a Nazi past affects the destination image of Linz, Austria, which has an identity deeply rooted in the Nazi era, as it is where Adolf Hitler grew up. Following a sequential mixed-method approach, Iordanova collected qualitative data from 88 participants, and quantitative data via a questionnaire distributed to 188 Austrian and 212 international visitors. The findings revealed that DMOs' attempts have been successful in establishing Linz in the minds of international visitors as a destination with an image dominated by contemporary art and culture, while for domestic tourists, some associations to its dark past still exist. The Nazi past appears to be still deeply rooted in the Linz brand, while actual experience with the destination positively modified its image. Local DMOs' strategy to incorporate the Nazi past into various exhibitions appeared to neutralize any negative connotations. The study contributes to the management and marketing of destinations with difficult heritage (e.g. crime, war, dark history), which are advised to embrace such pasts serenely in their brand strategies in a mature, facts-based manner, aimed at addressing previous stereotypes and turning them into a positive educational experience.

Chapter 10, by Carina Ren and Elizabeth Cooper, explored how users (social media influencers) negotiate online presence and social media content during their travels in Greenland and Denmark within a context of geographical peripherally, and a place where digital connectivity is limited and expensive. Data were drawn from five interviews conducted with professional social media influencers who visited Greenland for five days, with the trip sponsored by the local DMO to promote the destination. The chapter offers unique insights into the trade-offs and balancing acts of being 'online' and 'off the beaten track', seeing them as equally important but often challenging to reconcile. The chapter concludes that the social media influencers highlighted three predominate practices: the blurring of work/leisure and personal/professional; the technological mediation; and the co-creation of the product. Another notable finding is that that efficient and stable connectivity is not always a necessary condition for travelling and sharing in remote destinations. These findings have implications for destination managers in 'off the beaten track' adventure destinations like Greenland.

In the final chapter, Chapter 11, Jordi Arcos-Pumarola and Marta Conill-Tetuà explore how the intangible cultural heritage can be promoted through social networks. The authors focus on the case study of the *Fête de l'Ours*, held annually in the Vallespir region, France, and examine images of the festival posted on Instagram by local residents, visitors, tourism professionals and official tourism bodies in an attempt to understand how these users share and explain their experiences of the festival. The authors acknowledge that intangible cultural heritage is increasingly becoming an attraction in tourism and a space for exchange and recognition of culture. Such exploitation of the intangible cultural heritage raises issues related to its authenticity and preservation. As social media nowadays play a critical role in the formation of the image of a destination, the risk of alterations in traditions to match tourists' needs, and a loss of residents' identification to their culture, is becoming evident. The study assists destination marketers in devising strategies that can be used to promote and transmit the meanings of intangible cultural heritage in the context of social networks, without comprising their authentic character.

Overall, this volume contributes to the documentation and understanding of tourism marketing practices in the ten western European countries and territories explored. The wider application of digital marketing has been substantiated both conceptually in the first chapter, but also in the case of British young travellers examined in Chapter 2, and among visitors in central

Portugal presented in Chapter 3. Another key theme that emerges is the need to diversify the tourism product, especially as some areas in western Europe have reached the phase of stagnation/decline and suffer from overcrowding and seasonality. This is becoming a priority especially in the post-Covid-19 era. The fourth chapter, focusing on Italy, is such an example, where the need for participatory planning in destination marketing is proposed. Spain, another western European country that has been significantly affected by the pandemic, has strong potential in developing forms of gastronomic tourism such as wine and olive oil, as presented in Chapter 5. The hybridity of the tourism product experienced in some destinations like the Alps pose also an opportunity and a challenge for destination marketing organizations, as discussed in Chapter 6. Another type of mild development and its potential in a well-visited destination like Greece was explored in Chapter 7. A unique segmentation base, that of food neophobia, was used in Chapter 8, helping to further diversify the market. The challenges of marketing destinations with a dark past were recognized in Chapter 9, while the issue of limited connectivity in remote destinations was presented in Chapter 10. Finally, the challenge of balancing promotion of intangible cultural heritage and its preservation was highlighted in Chapter 11. Altogether, several of the above-mentioned cases explore the prominent role of digital marketing, with its successful implementation, however, depending on a number of factors to be considered by DMOs and other destination actors.

While the book offers insights into tourism marketing in western European countries, further research is necessary. Although a good representation was achieved by considering ten key countries and territories in this sub-region, the complexity and size of countries like Spain, Germany, the UK, France and Italy call for additional contributions from these areas. Somewhat similarly, other countries of this region like The Netherlands, Belgium and Ireland should be considered in the future.

Lastly, despite the wide coverage of topics, particular marketing issues in each country might have been neglected, as it was not possible to identify and include all of them in a single volume. Key lessons, however, can be drawn that are useful for country, region and city destination marketers, managers and policymakers.

Index

Note: The page numbers in bold and italics and with letter n represents tables, figures and note numbers respectively.